ARCTIC ENVIRONMENTAL COOPERATION

Arctic Environmental Cooperation
A study in governmentality

MONICA TENNBERG

Routledge
Taylor & Francis Group

LONDON AND NEW YORK

First published 2000 by Ashgate Publishing

Reissued 2018 by Routledge
2 Park Square, Milton Park, Abingdon, Oxon, OX14 4RN
711 Third Avenue, New York, NY 10017, USA

Routledge is an imprint of the Taylor & Francis Group, an informa business

Publisher's Note
The publisher has gone to great lengths to ensure the quality of this reprint but points out that some imperfections in the original copies may be apparent.

Disclaimer
The publisher has made every effort to trace copyright holders and welcomes correspondence from those they have been unable to contact.

A Library of Congress record exists under LC control number: 99076155

ISBN 13: 978-1-138-72014-5 (hbk)
ISBN 13: 978-1-138-72010-7 (pbk)
ISBN 13: 978-1-315-19517-9 (ebk)

Contents

Preface

This book is an edited version of my dissertation 'The Arctic Council. A Study in Governmentality' (University of Lapland, 1998). I am grateful to professor Harto Hakovirta who insisted that I do 'my own thing'. I am also grateful to Juha Auvinen, Vilho Harle, Helena Rytövuori-Apunen and David Scrivener for their comments during the final phase of this work. Clive Archer, Jyrki Käkönen and Oran Young also had an important role in helping to finish this story finally. I am, however, responsible for any shortcomings and mistakes that may appear in this work.

I have had the privilege of participating in the activities of the International Arctic Social Sciences Association (IASSA), first as treasurer and later as a member of the IASSA Council. This study would have been very different without this experience. I'm grateful to Linna and Ludger Müller-Wille for their friendship and cooperation.

Discussions with Tarja Saarelainen, Seija Tuulentie and Timo Koivurova on Michel Foucault, ethnic relations and international law at the University of Lapland have been important in advancing the intellectual process. This story would have remained untold without the opportunity to spend six months at the Institute of Arctic Studies, Dartmouth College in the USA. A grant from the Academy of Finland made this visit possible.

I am also grateful to the library staff at the University of Lapland, especially the interlibrary service, for their assistance over the past few years. To Lassi Heininen, Anne Nuorgam, Leif Rantala, Bruce Rigby, Oran Young and Henrik Elling I owe thanks for their help in collating the research material. The interviewees gave both their time and ideas to this study. For this I am deeply grateful.

Rovaniemi, 27 August, 1999

List of Abbreviations

AEPS	Arctic Environmental Protection Strategy
AMAP	Arctic Monitoring and Assessment Program
AMEC	Arctic Military Environmental Cooperation
ANCSA	Alaska Native Claims Settlement Act
ASDI	Arctic Sustainable Development Initiative
CAFF	Conservation of Arctic Flora and Fauna
CARC	Canadian Arctic Resources Council
ECE	Economic Commission of Europe, United Nations
EPPR	Emergency Prevention, Preparedness and Response
HOD	Head of Delegation
IASC	International Arctic Science Committee
IASSA	International Arctic Social Sciences Association
ICC	Inuit Circumpolar Conference
ICES	International Council for the Exploration of the Seas
ILO	International Labour Organization

IPO	Indigenous Peoples Organization
IPS	Indigenous Peoples' Secretariat
IRCS	Inuit Regional Conservation Strategy
IUCH	International Union on Circumpolar Health
IUCN	International Union for Conservation of Nature and Natural Resources
MMPA	Marine Mammal Protection Act
NGO	Non-Governmental Organization
PAME	Protection of the Arctic Marine Environment
SAAO	Senior Arctic Affairs' Official
SAO	Senior Arctic Official
TEK	Traditional Ecological Knowledge
TFSDU	Task Force on Sustainable Development and Utilization of Natural Resources
UNCED	United Nations' Conference on Environment and Development
UNCHE	United Nations' Conference on Human Environment
UNEP	United Nations' Environmental Program
UNGASS	United Nations' General Assembly Special Session
WWF	World Wide Fund for Nature

Major Events in the Arctic Cooperation

1987, 1 October	Gorbachev Speech in Murmansk, USSR
1989, 24 November	Mulroney Speech in Leningrad, USSR
1988, 12-15 December	Leningrad Conference on Scientific Research in the Arctic, USSR
1989, 20-26 September	AEPS Preparatory Meeting, Rovaniemi, Finland
1990, 18-23 April	AEPS Preparatory Meeting, Yellowknife, Canada
1990, 23 August	International Association of Arctic Social Sciences (IASSA) established
1990, 28 August	International Arctic Science Committee (IASC) established
1991, 15-17 January	AEPS Preparatory Meeting, Kiruna, Sweden
1991, 13-14 June	AEPS Ministerial Meeting, Rovaniemi, Finland
1991, 17-20 June	Arctic Leaders Summit, Horsholm, Denmark
1990, 8 November	Northern Forum established

1992, 3-14 June	UN Conference on Environment and Development, Rio de Janeiro, Brazil
1993, 11 January	Kirkenes Declaration on Barents Euro-Arctic Cooperation, Norway
1993, 16-17 August	Parliamentarians Conference, Reykjavik, Iceland
1993, 16 September	AEPS Ministerial Meeting, Nuuk, Greenland
1994, 15 June	Barents Euro-Arctic Region, Environmental Ministers' Meeting, Bodø, Norway
1996, 13-14 March	Parliamentarians Conference, Yellowknife, Canada
1996, 20-21 March	AEPS Ministerial Meeting, Inuvik, Canada
1996, 19 September	Establishment of the Arctic Council, Ottawa, Canada
1997, 12-13 June	AEPS Meeting, Alta, Norway
1997, 23-27 June	UN General Assembly Special Session, New York, United States of America

1 Introduction

Sustainable Development in the Arctic

A follow-up meeting to the United Nations' Conference on Environment and Development (UNCED) was held in June 1997 to evaluate the progress made towards sustainable development since 1992. The different Arctic states at the meeting voiced concerns regarding global threats to the northern environment. The Canadian prime minister Jean Chretien (1997, p.2) stressed the interdependent nature of environmental problems:

> Toxic chemicals do not respect borders. They even travel from distant sources to contaminate arctic food chains. This kind of threat can only be fought through international cooperation.

Circumpolar and global developments are also connected at an institutional level. Michael Grubb (1993, p.40) claims that 'almost all international institutions are potentially affected by the implications of sustainable development and the UNCED agenda' in the evaluation of the UN led process to advance a sustainable development world wide. However, efforts to define and enhance sustainable development are many. This is also the case in the Arctic.

The Arctic Council was established as 'a high-level forum' for cooperation on common Arctic issues, including environmental protection in the fall of 1996. International environmental cooperation in the circumpolar region was proposed almost ten years earlier, in 1987. Soviet president Mikhail Gorbachev (1997, p.22-23) suggested cooperation in the exploitation of the northern natural resources and in developing a comprehensive plan for the protection of the Arctic environment. Two years of preparatory meetings initiated by the Finnish government led to a ministerial conference in 1991 in Rovaniemi, Finland. Here, a Declaration on the Protection of the Arctic Environment and an Action Programme, known as the Arctic Environmental Protection Strategy (AEPS), was agreed. The eight Arctic states are Sweden, Denmark, Iceland, Norway, Finland, Russia (then the USSR), Canada and the United States of America.

The proposal for the Arctic Council was presented first at the AEPS ministerial meeting in Rovaniemi in 1991 but the idea did not develop. This was mainly due to the negative attitude of the U.S. government. The new Arctic Policy of the United States of America in 1994 provided the basis for a U.S. engagement in Arctic international forums and the strengthening of institutions of cooperation. Negotiations to establish the Arctic Council could be started the next year, following the meeting of U.S. President Bill Clinton and Canadian Prime Minister Jean Chretien in Canada. Negotiations were not easy; the establishment of the council was the result of a process in which concerns relating to the content, structure, and participation in the work of the Arctic Council was thoroughly discussed. Defining sustainable development and applying the idea in practice was one main disagreement among participants during the negotiations.

Theoretical Approaches in Research

Several writers have followed developments in Arctic cooperation. Three theoretical approaches in this study on cooperation have emerged during the last few years. They are based on:

1. the concept of security, particularly focusing on the role of natural resources and the environment in security thinking,
2. on different theories about regions, regionalism and regionalisation, and
3. approaches deriving from regime theory, emphasizing either the legal content or political significance of international cooperative efforts.

In contemplating the development of cooperation in the northern regions, the relationship between security considerations and natural resources is often considered more of a hindrance, rather than a source of cooperation (Möttölä, 1988; Dosman, 1989; Huitfeldt, Ries and Øyna, 1992). The importance of natural resources in security issues has been discussed by Archer (1988), Archer and Scrivener (1989) and Bergesen, Moe and Østreng (1987). Concern for the environment, however, has been a common feature for all the Arctic states. This has formed a new basis for cooperation since the late 1980's, which has led to a new combination of security notions and environmental concerns (Heininen, 1991). The importance of collective environmental security has been discussed in the Barents Euro-Arctic cooperation (Bröms, Eriksson and Svensson, 1994; Bröms 1995, 1997).

Studies in environmental security raise an interesting question about how the environment is defined in international relations theory. The words 'environmental' and 'ecological' are often used interchangeably. Richard Langlais

(1995, p.268) makes a difference between these two words when security is involved:

> Where the environmental implies a sense of surround, of surroundings, with the observer sharply defined as separate from the surroundings, the ecological, or, more precisely, the human ecological, denotes involvement that whenever possible includes the observer in the field that is being considered.

There is a fundamental difference in their use. This difference is about how the relationship between the actor and his or her environment is defined. Peter Bröms notes that 'the environment cannot be studied except through the understanding of the people'. The environment cannot be understood more than in 'social terms'. It cannot be comprehended solely as, an 'objective, real restraint' without regard to what the actors may think (Bröms, 1995, p.41).

The main question for regional studies is whether the Arctic has an identity as internationally meaningful region. In 1988, Franklyn Griffiths claimed that the Arctic was in a phase of transition, from a minimal political region with very little cooperation across borders to a 'coordinated region' where there are some efforts to establish cooperation (Griffiths, 1988, p.11). Young claimed in 1993 that the Arctic is emerging as a distinct region through current processes of cooperation (Young, 1993a, p.4). The development of Euro-Arctic Barents cooperation has increased interest among researchers into the problems of regionalism and regionalization (Stokke and Tunander, 1994; Dellenbrant and Olsson, 1994).

The process of region building is shaped into a 'bottom-up' process ('regionalism') as opposed to a process from the 'top-down' ('regionalization'). Regionalism is said to include spontaneity and a new grass-root level in the processes of cooperation, including intensively into cooperation actors other than states. Regionalization, on the other hand, is seen as an important strategy for peripheral states in strengthening the position of their relationships with centers and to defend that position in the world economy. The current processes of cooperation have been evaluated as far from the ideals of regionalism. The ideal would be for a regional actor to emerge in the north. This, however, is considered very unlikely (see, for example, Käkönen, 1996). It is more likely that region building serves other interests and the interests of actors other than local. Environmental cooperation in the Barents Euro-Arctic region, for example, is considered a part of Europeanization processes (Nilson, 1996).

Regionality is about framing problems and solutions about a distinct geographical region. Environmental concerns are also framed in regional terms. Framing implies that both political decision makers and the public find it 'natural' to address certain matters on a regional level (Castberg, Stokke and

Østreng, 1994, p. 72-75). In emphasizing the discursive side of regionality, Iver B. Neumann (1994, p.59) suggests that 'regions are talked and written to existence'.

From this perspective, studying regions becomes a project of analyzing the processes by which identities are created and evolved. This includes studying self-images and discourses and understanding the processes by which particular identities are shaped by histories, cultures, domestic factors and the ongoing processes of interaction with other regions (Hurrell, 1995, p.352-353). The relevant question in region building is the role the environment (and the concern for the environment) plays in regional identity building and the interaction among different actors.

The regime approach is particularly well established in the study of international environmental cooperation (Brown et al., 1977; Young, 1989; Gehring, 1994; Vogler, 1996a). Regimes can be defined as 'arrangements for institutionalized collaboration on topics and issues characterized by complex interdependence' (Haas, 1980, p.358). They can also be defined as, 'social institutions composed of agreed upon principles, norms, rules and decision-making procedures that govern the interactions of actors in specific issue areas' (Osherenko and Young, 1993, p.1). The regime theory has survived till the present, despite attacks on its being a woolly concept to which people apply different meanings (Strange, 1982), for its use of economic analogies and ignoring the context of cooperation (Walker, 1989; O'Meara, 1984), for repeating the problems of the state-centric approach (Milner, 1992) and for its claims of the false promise of international institutions (Mersheimer, 1994/95).

As Levy and colleagues (1995, p.267) note, the regime theory is 'alive and well'. There is a body of work on regimes and the development of the regime theory based on neoliberalism. This could be labeled as 'mainstream regime theory' or 'standard regime theory,' or as Robert O. Keohane (1988, p.381) calls the 'rationalistic study' of international institutions.

Constructivist, or reflectivist thinking, has challenged the standard regime theory's ontological and epistemological basis (Kratochwil and Ruggie, 1986; Czempiel and Rosenau, 1989; Rosenau and Czempiel, 1992; Behnke, 1993; Caporaso, 1993; Milner, 1993; Hurrell, 1993; Finnemore, 1996). The puzzle for a constructivist is how the human subjects constitute a social world, which in turn comprises the possible conditions for the actions of those subjects (Wendt, 1987, p.359). In a recent article Marlene Wind questioned whether the regime theory is worth saving from the constructivist critique of the rationalist, voluntarist and individualist assumptions of mainstream regime theory. To her, 'rationalist regime theory is and remains... a dead-end with severe and up until now unresolved ontological and epistemological inconsistencies' (Wind, 1997, p.258). My answer is positive 'yes'. The regime theory is worth saving. The current theoretical discussion provides ideas for rethinking the regime theory. Regimes are what 'we' - as students of international relations - decide to make of as concepts, research objects and theories about the nature of international relations.

One of the most important sources of influence in this book had been a short concise article by James Keeley on Foucaultian regime analysis (1990). This led me to study the works of Foucault. However, this is not a work on Foucault's thinking but one of using Foucaultian ideas to develop a regime theory restricted only by my limited understanding of Foucault. The point in this work is to reflect on the relations of power, knowledge and regime-building following Foucaultian ideas. That regime theory and power relations need to be more thoroughly analyzed is suggested by a recent evaluation of the state of the art in regime study by Hasenclever and colleagues (1996, p.205).

Foucault's ideas question the mainstream understanding of power relations in international relations theory and in particularly the idea that power considerations could, somehow, be external to a regime. This means that regimes are embedded in the networks of power. Foucaultian ideas challenge the macro-level analysis of power relations suggested by international relations theorists. Rather, power should be studied as events, in different space-time forms of its emergence, at the micro-level. Power is not only restrictive, but productive. Foucaultian politics are not those of the critical theorist attacking the visible world of power relations. Power is not external to human activities and interaction. Power is present in all human activities. All human activities take place in networks of power (Foucault, 1980a, p.98).

Foucault provides a very special view on the processes of institutionalization. Institutionalization is a process of power; institutions are concentrations of power. Institutions govern human relations. To govern 'is to structure the possible field of action of others' (Foucault, 1982a, p.221). Governmentality is a question of analyzing 'a 'regime of practices' - practices being understood here as places where what is said and what is done, rules imposed and reasons given, the planned and the taken for granted meet and interconnect' (Foucault, 1991a, p.75). A study on governmentality is focused, not to the institutions and the process of institutionalization as such, but on the practices of power created and maintained through institutionalization.

The Research Problem

One may ask why there is a need for environmental cooperation since there are already several agreements covering the Arctic. Three explanations have been given. First, the existing agreements have been evaluated as inadequate because the existing arrangements do not address the specific needs of environmental protection and the regional conditions. Second, some existing arrangements have been evaluated as narrow and shallow. There is a problem of coverage for existing mechanisms; not they have signed or ratified all the agreements by all the states in the region. Finally, it has been claimed that not enough attention has been paid to the existing mechanisms of environmental interdependence. Many environmental issues and problems that have a transboundary nature are not handled within the current environmental protection regime. The problem is that the Arctic environment is in need of more extensive protection than individual states or existing various legal arrangements can provide (Stokke, 1990; Hoel et al., 1993; Rothwell, 1994).

The present state of Arctic environmental cooperation does not seem really to tackle the identified problems of the Arctic environment and its protection. Current cooperation does not produce the new agreements or binding conventions for the protection of the Arctic environment. These would deal with the application of the agreements or their existing problems. The cooperation is not creating new mechanisms for helping Arctic states to comply with existing arrangements. What is really needed are an implementation, and some consideration of the special situation and requirements of the Arctic. The critical issue is the financial and technological help for the cooperative parties that have problems in implementing existing agreements, particularly considering the special conditions of the Arctic.

None of these aspects, however, are part of the existing cooperative arrangements. Instead, cooperation so far has produced recommendations, action programs and guidelines that do not necessarily improve the quality of the environment or ensure more efficient measures to protect the environment from new threats in the region. Thus, the question is what is happening in the Arctic concerning international environmental cooperation. To study this problem, this research aims at answering three questions:

1. what is the meaning of environmental cooperation in the Arctic?
2. what does the establishment of the Arctic Council and inclusion of AEPS activities to its mandate mean for international environmental cooperation and the protection of the environment in the Arctic? And
3. what does studying the Arctic case give for the development of regime theory and research?

On a very superficial level the meaning of cooperation is, of course, 'to protect the environment'. Different people attach a variety of meanings to this idea, and these reflect varying sets of human values in relation to the environment. Regimes are packed with meanings, which are under continuous negotiation and renegotiation through processes of definition and redefinition. Social institutions, such as regimes, and their purposes are associated with particular meanings (Conca, 1994, p.11-12).

The question about the meaning of cooperation has to be directed to the representatives of the different participants, not only to the diplomats and government officials. Although individuals were interviewed, the focus is not on individual meanings but in intersubjective meanings created in the cooperation. Intersubjectivity stresses that meanings are for subjects but they are not produced in a vacuum; they are produced for a subject or a group of subjects. Meanings are not separate from each other. Meanings cannot be identified except in relation to others. Those meanings are collective, that is, intersubjective (see Taylor, 1987; Fay, 1987). With an environmental protection regime, the struggle is over the meanings attached to the human-environment relationship. 'Imposing a meaning' suggests that regimes could also be seen as arenas for conflict and the exercise of power. From this perspective, regimes are foci and loci of struggles in meanings produced by different participants. Regimes not only react to a configuration of power but also to a configuration of dominant social purposes (Keeley, 1990, p.95).

The second question - the meaning of institutionalizing international environmental cooperation - refers to the reason for establishing the Arctic Council. This question cannot be understood without taking a closer look at the role of regimes in international relations. For the mainstream regime theorists, governance is an effort to manage two systems, the natural and the human, and means to control the increasing problems of interaction between these systems (see Choucri, 1993; Young and Druckman, 1992). Efforts at environmental governance include creating systems aimed at reconciling the conflicting interests of the actors while minimizing human-induced disturbances of the natural system (Young, 1993a, p.6).

The mainstream regime approach is agnostic about the changes in the human-environment relationship even when the studies focus on environmental protection regimes. According to this view, 'truly effective international environmental institutions would improve the quality of the global environment...' (Keohane, Haas and Levy, 1993, p.7). Because of the short-time of cooperation, such impacts are difficult to detect. The focus is on behavioral change, not an improvement in the quality of the environment.

One might ask what is the point in studying international environmental cooperation without at least a minimal interest in the effects of cooperation on human-environment relationship. Even if changes in the quality of the environment cannot be detected right now, changes in the meanings attached to the environment can be noticed. Concentrating on the 'mentalities' of governing rather than the effectiveness of the regulatory processes means studying ways of understanding the world and being in it. Studying mentalities is an effort to capture changes in collective ways of being; that is to study the collective meanings, norms, attitudes, knowledge, conventions, and the ways to perceive the world and respond to it. The word 'mentality' refers to the slow-changing collective understandings of human existence and relations with the environment (see Peltonen, 1992, p.15).

For the mainstream regime theorists, governance is 'any purposeful activity intended to 'control' or influence someone else that either occurs in the arena occupied by nations or occurring at other levels, projects influence into that arena' (Finkelstein, 1995, p.368). Regimes decrease states' vulnerability from independent action and reduce uncertainty stemming from uncoordinated activity. States maintain some degree of control over each other's behavior through regimes (Keohane, 1982, p.351).

For those who see international relations more as a society than as a system the problem of governing is not of control but of constitution. Governmentality is about creating and maintaining order. By emphasizing that international regimes are 'not simply by some descriptive inventory of their concrete elements, but their generative grammar, underlying principles of order and meaning that shape the manner of their formation and transformation,' (Ruggie, 1982, p.380) the constitutive quality of regimes is emphasized. For Lynton Keith Caldwell, international environmental diplomacy over the last two decades has produced a 'new international order' in the field of international environmental politics. There are institutions, regimes, treaties, nonbinding guidelines and financial mechanisms to make this order (Caldwell, 1990a, p.128).

Finally, Arctic cases of environmental cooperation have also been used earlier to discuss some basic assumptions of the regime theory. For example, the study on Arctic environmental protection regimes by Young and Osherenko (1993) aims to develop a multivariate model for the study of a regime. This point questions the tradition in the Arctic of emphasizing the exotic and unique features of the region. Of course, the negotiation process itself is unique but emphasizing the exotic character of Arctic developments results in setting the region apart from the concerns of mainstream study (see Young, 1992, p.13).

The Strategy and Structure of Study

The debate on the problems of the international relations theory overall has accompanied attempts to produce empirical knowledge on international regimes. Studying governmentality not only means questioning the ontological status of international regimes, but also the methods of investigating them. An interpretive approach can be used to gain new insights into the constitution and functioning of regulatory international institutions. An interpretive approach not only offers the means for understanding the self-interpretation and self-definition of human collectivities. It can also generate insights into the very orders in which regulatory international institutions are embedded. Both institutions and orders are constituted by intersubjective meanings. Regulatory institutions and their underlying orders consist of social practices (Neufeld, 1995, p.90).

As a methodological tool a textually oriented discourse analysis is used in this work. The research is based on a textual analysis of documents produced in negotiations and the answers given in interviews and documents. Foucaultian discourse analysis, that is interpretive analytics, claims that a discourse is not only a statement but also a statement connected to social practice. A textually oriented discourse analysis treats discourse three-dimensionally and aims to find the relationship between text, discourse and practice; any discursive event is seen as 'being simultaneously a piece of text, an instance of discursive practice, and an instance of social practice' (Fairclough, 1992, p.4). Therefore, even if texts have been studied, the interest is not in the texts as literary works but in the meanings and practices of governing that can be traced into those texts (Dryzek, 1997, p.76).

Talk about the environment within a regime is an example of 'discursive diplomacy' (Wettestad, 1994). Studying this talk directs attention to the main content of international environmental diplomacy (verbal acts as speeches presented and arguments exchanged over the negotiation table) and to the discourse of environmental diplomacy (see Austin, 1962; Searle, 1969 about speech acts). Discourse is that 'which is produced (perhaps all that was produced) by groups of signs' (Foucault, 1972, p.107). In the discourse, the objects and subjects of international environmental politics are defined by using different concepts, enunciative modalities and strategies that constitute the human-environment relationship. How concern for the environment is constructed in diplomatic discourse is not only talk. Discourses do not reflect or represent social entities and relations, they construct or constitute them. When embodied in practices, a discourse becomes a creative part of the reality it purports to understand (Woolin, 1988, p.184).

Making sense of Arctic developments requires understanding and interpretation of meanings and practices of cooperation. Understanding aims to give an 'insider view' of the world as it is experienced by the actors, and provide an account of what constitutes meaningful action. This method directs attention to the practices and self-understandings of the actors. The practices of cooperation and their effects are analyzed, not as to legal rules (this would be the approach by the new institutionalists), but for meanings and changes over the meanings on the human-environment relationship. The aim of interpretation is not to say what is wrong or what is right. The focus is on:

> ... pointing out what kind of assumptions, what kind familiar, unchallenged, unconsidered modes of thought the practice that we accept rest (Foucault, 1988a, p.154).

Interpretive analytics are based on the idea that the researcher and the researched are assumed to be part and parcel of a single process. The researcher and the research object share the same world, although they do not have same experiences of it (Hollis and Smith, 1991, p.72-73). However, only a researcher can give meaning to research itself, which they connect to the chosen themes of the research. Studying the 'mentality of governing' means analyzing the effects of power; what makes some form of activity thinkable and practicable both to its practitioners and to those upon whom it is practiced. Governmentality asks what the purpose of power is and how it works. Political rationality, however neutral it may seem, includes purposive or value rationality. Political rationalities conceptualize and justify goals as well as the means to them (Gordon, 1991, p.3).

Discourse about the Arctic environment and its future can be found in the documents produced in ministerial meetings, the meetings of senior Arctic affairs officials, and in the documents and papers of working groups. These texts are relevant, since it is possible to find the objectified world 'in common' in them. These texts coordinate the acts, decisions, policies and plans of the actual subjects. In addition, interviews have produced additional material that, in the main, supports the search for what was essential in diplomatic environmental discourse.

The AEPS and the Arctic Council negotiations have produced a formidable amount of paper in such a short time. The interviews helped to determine the main themes of the discussion. They were open-ended and centered around four questions: the problem of the human-environment relationship in the Arctic, the issue of participation in cooperation, the meaning of knowledge in the process and the point of cooperation in the Arctic. Interviews were continued until they became repetitive; until no new themes or issues were raised. Most of the interviews were held while the negotiations to establish the Arctic Council continued. Some follow-up discussions were conducted after the interviews.

The material used in this study is not perfect and collecting it was not a simple task. The record-keeping for the negotiation process after 1991 has not been continuous. It forms no logic, at least not one that I have not discovered, and the host countries and their practices differ. Therefore, for example, material obtained from the Nuuk meeting in 1993 is not complete. I had to rely on individuals to provide the material for me. I also noticed differences, probably national ones, in the openness and ease of the process of obtaining drafts and documents. Another problem was with the material from the Russian side. Because of my lack of linguistic skills and the lack of response to my inquiries by the Russians, the material leaves room for improvement. Despite these inadequacies, with the help of interviews there was enough material available to reconstruct developments.

The structure of this work is based on the texts and issues that arise from them. Three power-related themes arise from the analyzed texts:

1. who has the power to take care of the Arctic, its environment and peoples - that is, how the role of the actors is constituted in the Arctic,
2. whose understanding of the Arctic environment and its state counts the most - that is, whether or not knowledge is power and vice versa, and
3. how the human-environment relationship is defined in the Arctic - that is, constructing this relationship is an act of power.

These themes are discussed in relation to regime theory and its assumptions. The special relationship between states and indigenous peoples is studied in Chapter 2. This relationship needs to be understood to make sense of the cooperative processes. Having knowledge of historical background is important, since the present can be only understood through knowing the past. The issue is the variety of claims on the environment by the sovereign states, indigenous claims to self-determination in environmental issues and their recognition from those states as relevant actors. In addition, even if the states are the recognized actors in international environmental politics, the nature of current environmental conditions challenges the range of action for the individual states. The question is whether Arctic states on their own, or in cooperation with each other, can actually deal with the problems at hand.

With environmental cooperation, knowledge of pollutants and anthropogenic emissions is important in the management of the problems but there is more to this than instrumental knowledge. The constitutive character of knowledge refers to different environmental world views and the various kinds of knowledge their carriers have. This aspect of cooperation is discussed in Chapter 3. The discourse on the environment and the ways it has formed itself into cooperation is studied. Issues, such as who knows what, are central to Arctic environmental cooperation. The relationship between scientists, Arctic organizations and the indigenous peoples and its effect on the content of the Arctic environmental cooperation is discussed.

The human-environment relationship can be found in different world views. These world views include aspects of ethical stances on relations between man, the environment and the world. Chapter 4 discusses the human-environment relationship in Arctic environmental cooperation and the changes in it over almost ten years of negotiations. This chapter looks into the internal dynamics of this discourse and the interests in the process defined by both states and others. An explanation for developments is not sought outside the negotiation process itself. Analysis of the reasons and explanations of the actors themselves is the focus of the study.

Finally, in Chapter 5, the transfer from the AEPS to the Arctic Council era is described and analyzed. The point in studying governmentality is not to tell the story how it 'really' was but to study how the authorities and the rationalities of governing have made the world as it is now understood (Simons, 1995, p.38). It is through discourses that make it seem as if the governmental techniques of addressing a problem are based on shared logic and principles. Studying governmentality challenges self-evident political rationalities, even for the protection of the environment. Foucaultian doubt about one rationality of governing opens the possibility of studying different ways and rationalities of governing.

2 Discourse on Sovereignty

The Need for Special Measures

State Sovereignty in the Arctic

In international practice, sovereignty is a claim about the way power is or should be exercised. Sovereignty is a rule that provides order in international relations; it is 'both theory and practice aimed at establishing order and clarity in an otherwise turbulent and incoherent world' (Camilleri and Falk, 1992, p. 1). Above all, sovereignty is an issue of identity and the constitution of identities. The theory and practice of state sovereignty formalize a specific answer to questions about who 'we' are as political beings. Sovereignty defines a social identity; its core is the notion of political authority as lying exclusively in the hands of spatially differentiated states (Walker and Mendlovitz, 1990; see also Bloom, 1990; Wendt, 1994).

The issue of sovereignty has been a part of the Arctic environmental cooperation from the beginning. The Swedish delegation reminded the participants of Principle 21 of the United Nations' Conference on the Human Environment, UNCHE (1972) at the first preparatory meeting on the protection of the Arctic environment in 1989. According to this principle, sovereignty over natural resources and the environment belongs to the state. According to this principle:

> States have, in accordance with the Charter of the United Nations and the principles of international law, the sovereign right to exploit their own resources pursuant to their own environmental policies, and the responsibility to ensure that activities within their jurisdiction or control do not cause damage to the environment of other states or of areas beyond the limits of national jurisdiction (A/CONF.48/14/Rev.1).

This principle states that the activities of one country must not be allowed to affect negatively the environment of other countries. The principle was reaffirmed in Rio de Janeiro in 1992 in the UNCED. 'Needless to say, this principle is applicable to the Arctic region as well' (Statement by the Swedish Delegation, 1989, p.3).

For Norway in particular, the issue of sovereignty seemed important at the first preparatory meeting of the AEPS:

> The challenges of securing safe and rational management of resources, protection of the environment and strategic stability are linked up also with questions of jurisdiction and sovereignty (Statement of the Norwegian Delegation, 1989, p.1).

The Norwegian position in these areas is determined by Norway's strategic location emphasising '... our role as the state having sovereignty over the Svalbard archipelago as well as Norway's jurisdiction over large sea and shelf areas' (Statement of the Norwegian Delegation, 1989, p.10-11).

In the negotiations to establish the Arctic Council, the dispute over the marine border between Canada and United States emerged for a short while, but this was put aside. Besides their boundary delimitation conflict, there are also other jurisdictional conflicts in the Arctic between Canada and the USA. The USA disputes the legal status of the northern waters in the Canadian Arctic. According to Canadian interpretation these waters are 'internal' but the Americans claim that the North West Passage is 'international' waters (Brelsford, 1996a).

The problem of state sovereignty over the environment and natural resources is emphasized by the awareness that there is a growing number of 'international' environmental problems. The international character of environmental problems is also evident in the Arctic: 'The pollution problem of today does not respect national boundaries' (AEPS, 1991, p.1). It follows that the limits of state sovereignty have to be discussed in dealing with the problems at hand in the Arctic.

Some consider state sovereignty as the main problem in the management of environmental problems (Camilleri and Falk, 1992, p.185-186). For example, the Norwegian representative pointed out that two relevant categories of the participating countries the states which have sovereignty and jurisdiction in these areas and other states whose nationals are engaged in activities in the same region. According to the Norwegian view: 'with regard to the High Arctic the group of states participating in this preparatory meeting is at one and the same time too broad and too limited'. It is 'too broad in relation to the issues concerning exercise of national jurisdiction and too limited with regard to issues concerning obligations on states active in the areas in question' (Statement of the Norwegian Delegation, 1989, p.14).

These comments by arctic states suggest the issue of sovereignty is a living concern for at least a part of the participants in the cooperation. Jens Bartelson (1995, p.45) suggests the rule of state sovereignty is 'a parergon' - a frame. A parergon does not exist in the same sense as that which it helps to constitute. There is a ceaseless activity of framing, but the frame itself is never present, since it is, itself, unframed. At times, however, the frame can be seen.

'Arctic' States

Different states used environmental concerns to construct their 'Arctic' identity. For some participating states the connection to the Arctic is obvious; 'Some are Arctic rim states with coast lines to the Arctic Ocean and with jurisdiction over land areas in the High North'. For others, the connection to the Arctic is not so obvious; 'Other have territory north of the Arctic circle which in climate and nature clearly differ from the region of the High North' (Statement of the Norwegian Delegation, 1989, p.14).

Most of the other participating countries had a clear Arctic identity and concern for their Arctic environment. According to the Russian view the 'Arctic region provided and provides a lot for our country'. The Russian representative emphasized the concern for the situation in the Arctic; 'its clear tendency to deteriorate is a cause for anxiety' (Address of the Soviet Representative, 1989, p.1). Norway presented herself as 'a coastal state to the Arctic Ocean the northern parts of Norway have from the oldest times been dependent of the living resources in the sea for its livelihood'. The focus of concern was the state of the marine environment; the marine ecobalance is 'essential for life and livelihood in Norwegian coastal regions in the North' (Statement of the Norwegian Delegation, 1989, p.11).

For the Danish, the identity of the country and participation was clear:

> Denmark is one of the eight Arctic nations that borders directly on the Polar Basin. Geographically Denmark is closer to the North Pole than any other country in the world, the distance from northernmost Greenland being a mere 460 miles (Statement by the Danish Delegation, 1989, p.1).

Icelandic representative considered their Arctic identity in relation to its resources; 'fishing areas around Iceland are of vital importance for her economy' (Gudnason, 1991, p.2). Marine pollution is the main cause for concern since 'Iceland is very sensitive to conditions in the Arctic, being situated in a gateway to the North Atlantic' (Iceland, 1990, p.1).

In particular, a difference in their Arctic identities can be found between Finnish and Canadian approaches. The difference is in whether the concern is for the well-being of the natural environment or for the well-being of the northern habitants. The basis for Finnish initiative was the concern for the impact of long-range transboundary pollution on the forests in eastern Lapland For Finland, 'in the long-run air pollutants constitute a serious threat to our forests and our forest economy' (Pietikäinen, 1991, p.3).

For Canadians, the Arctic is seen as the home of indigenous peoples; 'the Arctic is a home to 73,000 native and nonnative Canadians. In other words, our Arctic comprises a rich, multicultural mosaic' (Campeau, 1990, p.1). The state and quality of the environment for these peoples are important; 'both Indians and Inuit have depended on the land as basis of their culture - they have relied on its resources for food, clothing and income' (Siddon, 1991, p. 2).

There were two countries which did not really indicate how they saw themselves as 'Arctic states:' Sweden and the United States. Both countries emphasized global concerns for the state of the Arctic environment. The Swedish representative mentioned the concern for global problems in the Arctic, such as the depletion of the ozone layer and climate change. For the Swedish, cooperation in the Arctic could be an example for the rest of the international community; 'Sharing our experience might serve as an inspiring example for other countries and regions' (Dahl, 1991, p.2).

For the United States, at least at the beginning of cooperation, the region is mainly interesting for research on global environmental problems:

The delicate balance of its physical, chemical and ecological components, governed by the very low rate of biogenesis and chemical turnover in large masses of freshwater and sea ice, makes the Arctic an 'early warning system' for global change, where the signatures of climate change are expected to occur first (Weinman, 1991, p.3).

For the United States, the Arctic most of all is of global significance as, 'an ecologically sensitive region' that provides 'livelihood for indigenous peoples; majestic scenery; splendid wildlife resources to be shared and managed cooperatively by a diversity of political jurisdictions; and a wealth of marine species and mineral deposits that benefit the rest of the world' (Weinman, 1991, p.1).

Protecting the Environment

States are important in defining and maintaining property rights. Daniel Bromley points out that property is 'a social relation between the benefit stream from the property, the holders of rights to that property and those who bear duties'. Property right is a claim to a benefit stream. According to this perspective, environmental problems - because they are usually a matter of the private interest of A vs. the private interest of B - can be regarded as triadic: A, B and the state (Bromley 1991, p.19). The state agrees to protect that right. Issues such as who will get the rights to land and natural resources, and who will have the protection of the state to do as they wish with those assets are important in environmental policies and management. From this standpoint, the environmental policy problem is 'nothing but a struggle about who shall have control over the stream of future environmental services' (Bromley 1991, p.38; Peluso, 1993; Young, 1993b).

The states recognized their role as major actors in Arctic environmental cooperation. The preparatory meeting in Rovaniemi 1989 resulted in affirming the concern for the Arctic environment and the need for special measures by the states to protect the environment. The Finnish initiative acknowledged the number of international, regional and bilateral agreements on the environment which apply to the Arctic. Nevertheless, according to the Finnish view, it is 'of utmost importance that a mechanism is created to complete the provisions of those disparate agreements so as to ensure their effective application'. The responsibilities of the states was emphasized; 'It is evident and necessary to tighten up national measures: political, legal, technical, or others taken so far in nearly all environmental sectors' (Statement by the Finnish Delegation, 1989, p.2).

The Finnish proposal for the continuation of cooperation was to develop a process which could lead to a program of action composed of concrete and practical measures. This could include: 'deepening of the understanding of the problems' by a lead country method, and 'assessment of the state of the environment and impacts of economic and social activities and exchange of information in case of pollution incidents'. Cooperation might further include 'accords on concrete measures' on most urgent problems such as pollution of the sea, acidification, accumulation of toxic chemicals and radioactive contamination. Finnish representatives stressed the need to develop a long term strategy and policy to improve the environment and prevent its further contamination. This could lead to either a political or legal agreement. The Finns emphasized the need of rapid action; 'the sooner we can research an agreement, political or legal, the better' (Statement by the Finnish Delegation, 1989, p.3). The Finnish initiative was open; the point was in starting cooperation and leaving the cooperation to take its form in further negotiations.

The Russians supported the idea of strong legal measures to protect the Arctic environment. The Russians recognized the need 'for blocking measures' against the deterioration of the environment (Address of the Soviet Representative, 1989, p.3). The need of legal measures was emphasized; the 'effective solution of the problem is possible only through broad and complex cooperation between the Northern countries, based on a solid and permanent legal ground' (Address of the Soviet Representative, 1989, p.7).

The Russian representative suggested 'a sort of code of civilized and environmentally sound conduct of states'. It would determine and provide for an 'organic interrelation' between their rights and obligations vis-à-vis nature and each other. This environmental code would cover all the participants and all the regions in the Arctic. It should be based on the recognition of 'the unconditioned rights of every person to live in the most favorable environment' (Statement from the Soviet Delegation, 1990, p.3).

The environmental code suggested by the Russian delegation included several principles. First, the environmental 'well-being' of any particular state can not be achieved at the expense of other states or without due account of their interests. Second, no activity should be detrimental to the environment, be it within or outside the framework of national jurisdiction. Third, any activities having unpredictable environmental consequences should be inadmissible, that is, putting the precautionary principle into practice. Fourth, 'free and unrestricted' exchange of scientific and technological information' on the problems of the environment and of modern environmentally sound technologies is required. Finally, coordinated measures should be effectively taken at different levels of action, including international, regional and national levels (Statement from the Soviet Delegation, 1990, p.3).

No delegation suggested that the existing system of legal measures was adequate. There was an agreement 'that issues are not covered by existing conventions, and a number of threats to the environment were not adequately dealt with today' (Statement by the Swedish Delegation, 1989, p.1). According to a report prepared in the preparatory process, only the agreement on polar bears and some individual provisions in other agreements were considered to address the Arctic region directly. The report of the consultative meeting concluded that there were a number of areas of environmental protection where the particular Arctic conditions should be more distinctly reflected. The existing legal instruments was seen as the basis for 'improved Arctic environmental protection through a strengthening and broader application' (Protecting the Arctic Environment, 1990, p.3).

Most of the participating states supported the idea of special legal measures to protect the environment in principle, but wanted to avoid overlapping. What was needed, according to the Swedish delegation, were 'clearer guidelines and measures to protect the environment'. Rules also had to be followed. The Swedish view was that much can be accomplished through existing conventions and agreements; a wide range of international treaties applicable to the Arctic region exist (Statement by the Swedish Delegation, 1989, p.3-4). The Canadians emphasized that the process of elaborating multilateral legal instruments can be 'arduous and time consuming,' so if 'we agree on the need for one or more such instruments we should begin the preparatory process as soon as possible'. The Canadians stressed that the negotiation of a treaty or other legal instrument is not enough: 'The treaty must be brought into force and many well founded treaties take years to come into force because of the slow pace of the ratification process'. They were concerned about delays in the 'light of the accelerating threats to the Arctic environment'. Moreover, it was not enough to ratify treaties. They have to be applied and implemented. The Canadians emphasized the need for 'concrete measures to apply and enforce the obligations we accept' (Statement by the Canadian Delegation, 1989, p.4). Despite these reservations the Canadians believed that an 'urgent need exists for comprehensive measures to safeguard our northern ecosystems from the adverse effects of human activities' (Statement by the Canadian Delegation, 1989, p.2).

Most critical to developing new regional and legal measures to protect the Arctic was Norway. The Norwegians stressed that 'we should avoid overlapping and double work with earlier and ongoing efforts'. They pointed out the importance of other international efforts; 'We shall be well advised to keep in mind also the numerous other ongoing initiatives dealing with environmental problems both at regional and global levels' (Statement of the Norwegian Delegation, 1989, p.15).

In addition, according to Norway, 'the legal framework applicable in the Arctic would be strengthened considerably if participating states would become parties to various legal instruments that are already in place' (Opening Statement by the Head of the Norwegian Delegation, 1990, p.4). There was no need for new rules:

> ... unless participating states would also become contracting parties to the basic instruments which already exist, chances are less that new rules would become really effective (Opening Statement by the Head of the Norwegian Delegation, 1990, p.5).

The aim of the Arctic environmental cooperation should, according to Norway, be to seek and identify problems which were not already covered by existing international agreements. The solution to the environmental challenges and problems in the Arctic cannot be sought in isolation from broader and regional efforts at protecting and improving the natural environment of man. The Norwegian view summarizes the approach chosen then; 'To a large extent, the relevant agreements are already in place, only waiting for wider adherence and more effective implementation' (Stoltenberg, 1991, p.2).

The AEPS Working Groups

The sense of the need for special measures in the Arctic region had, to a large extent, been lost in the preparatory process by 1991. The political interpretation of the preparatory process was actually that few special measures were needed. The countries aimed at a 'practical' and 'pragmatic' approach; according to the statement by the chair of the Yellowknife meeting, 'it seems clear that the process of Arctic environmental cooperation is evolving into a practical and pragmatic search for solutions to issues of common interest' (Statement by the Chair, 1990, p.2).

The concrete measure taken by the states in 1991 was the establishment of four different working groups: Arctic Monitoring and Assessment Program (AMAP), Conservation of Arctic Flora and Fauna (CAFF), Protection of the Arctic Marine Environment (PAME) and Emergency Prevention, Preparedness and Response (EPPR).

The object of the AMAP was the measurement of pollution levels in the Arctic environment and their effect assessment. The states noted that the pollution data available from the region was mostly based on national research programs. In order to have better documentation on the environmental situation in the Arctic international cooperation was needed (AEPS, 1991, p.31). AMAP was established: to provide 'integrated assessment reports on status and trends in the condition of Arctic ecosystems,'to identify 'possible causes for changing conditions, 'to detect 'emerging problems, their possible cause, and the potential risk to Arctic ecosystems including indigenous peoples and other Arctic residents,' and to 'recommend actions required to reduce risks to Arctic ecosystems' (AEPS, 1991, p.33).

CAFF was established to become 'a distinct forum for scientists, indigenous peoples and conservation managers' in the Arctic to exchange data and information and to collaborate for more effective research, sustainable utilization and conservation (AEPS, 1991, p.40). Jeanne Pagnan from the CAFF secretariat describes the importance of the establishment of the working group on conservation; 'nothing as such existed before at all' (Pagnan, 1996).

Both PAME and EPPR became forums for state officials and can be compared with the other working groups which had larger participation. PAME was given the task of reviewing the relevance of international instruments for the protection of the marine environment in the Arctic (AEPS, 1991, p.34-35). EPPR was established to review existing bilateral and multilateral arrangements in order to evaluate the adequacy of the geographical region by cooperative arrangements (AEPS, 1991, p.37). These working groups have produced reports, guidelines and strategies. For example, PAME suggested the development of an Arctic regional action programme to address land-based sources of marine pollution and guidelines for offshore petroleum activities. The rationale for these guidelines, according to the PAME, is that no single instrument completely addresses the problems associated with land-based sources of marine pollution in the Arctic (PAME, 1996, p.14-16). The work of CAFF is another example of the development strategies and applications typical of the work done within the AEPS in dealing with regional concerns. The Habitat Conservation Strategy and the Circumpolar Protected Area Network are considered by CAFF to provide a common framework for the Arctic countries to ensure a necessary level of habitat protection. The species-based initiatives of CAFF contribute to the achievement of habitat and ecosystems-oriented goals, to the maintenance of the biodiversity within the Arctic regions, and provide the information needed for effective conservation and management actions (CAFF, 1995-1996, p.1-4).

Several guidelines were accepted by ministers at the AEPS meeting in Alta 1997: Guidelines for Environmental Impact Assessment in the Arctic, Arctic Offshore Oil and Gas Guidelines, and the Arctic Guide for Emergency Prevention, Preparedness and Response (SAAO, 1997). This approach was claimed to save time and money. Although the guidelines are not legally binding, they are still expected to make a difference to the current situation which does not have any direction. The strength of this approach is in avoiding duplication and increasing the awareness of the Arctic outside the region (Mähönen, 1996).

In a recent study on environmental protection in polar regions and international law, Donald Rothwell (1997, p.240) defined the AEPS as an example of 'soft' international law. Cooperation among Arctic states is based on compiling knowledge and developing action programmes, guidelines and strategies instead of legally binding international treaties. Rothwell (1995, p.281) describes the current Arctic environmental protection regime as, a 'collection of customary international law, fragmented multilateral and bilateral legal instruments dealing with some Arctic issues and global international instruments that have an impact in the Arctic'.

Most often regimes are defined as multilateral agreements among states which aim to regulate national actions within an issue area (Haggard and Simmons, 1987, p.495). For Haas (1980, p.358), they are 'norms, rules and procedures agreed to in order to regulate an issue-area'. The expected result of a regime is rules for dealing with the problem at hand. As Young (1980, p.333) suggests, 'the core of every international regime is a collection of rights and rules'. For the mainstream regime theorists, the rules written as agreements and conventions are thought of as 'specific prescriptions and proscriptions for action'. Norms, rules and principles are, in fact, considered by most students of the subject to be the 'basic defining characteristic of a regime'. This emphasis is seen in the definition of regimes developed by Stephen Krasner. For him, regimes are 'implicit or explicit principles, norms, rules and decision-making procedures around which actors' expectations converge in a given area of international relations' (Krasner, 1982, p.186).

Sovereignty, legal rules and the state require the other in a tight, unbroken circle; it requires that a formally connected set of rules operate exclusively in some manifold space and time. For a given time and a place there can be only one set of rules operating (Onuf, 1989, p.141). The term 'sovereignty' had, for a long time, expressed the idea that there is a final and absolute authority in the political community. Hinsley (1986, p.26) points out that this definition needs an addition 'and no final and absolute power exists elsewhere'. The enforcement assumed in the emphasis on the legality of rules is often 'an illusion,' for a constructivist. The fact that rules are not followed does not mean that there are no rules. Nor does the lack of international authority to enforce those rules make it impossible to have rules of action. According to Nicholas Onuf (1989, p.76), rule does not mean legal rule in the narrow sense of formal and enforceable.

Rules for the constructivist rules are persuasive to the extent that provide instrumental guidance and reflect moral considerations. Understanding norms as a rule and as embedded in social institutions act like structures shaping the behavior of states. The strength of this understanding is that such a norm as a rule must have an 'aura of legitimacy' despite the origin of the norm or the way it originated. The reason for following such norms is not based on enforcement but on their perceived legitimacy (Florini, 1996, p.364-365). This view is shared by Audie Klotz; international norms do not, strictly speaking, determine behavior since they constitute identities and interests, and define a range of legitimate policy options (Klotz, 1995, p.461-462; see also Cortell and Davis, 1996).

The most broad definition of a regime is that 'a regime exists in every substantive issue-area in international relations... wherever there is regularity in behavior, some kinds of principles, norms or rules must exist to account for it' (Puchala and Hopkins, 1982, p.246). Instead of strict rules and norms, according to Kratochwil and Ruggie (1986, p.764), 'we know regimes by their principled and shared understandings of desirable and acceptable forms of social behavior'. Puchala and Hopkins (1982, p.246-247) define regimes as intersubjective in that they exist primarily as 'participants understandings, expectations or convictions about legitimate, appropriate or moral behavior'. The rule of state sovereignty

... has conquered the world for the people by legitimizing the states, and only the states which claim to speak in their name, and it has elevated and institutionalized the progressive view of human affairs by attempting to freeze the political map in a way which has never previously been attempted (Mayall, 1990, p.56).

Foucault suggests that the rule of sovereignty collapse into practices of power. Starting from practices means concentrating on concrete events; practices are instances of people doing or saying or writing (Fairclough, 1992, p.57; Lemert and Gillan, 1982, p.34-38). These theoretical discussions point to an approach in which the connection of law and language is further studied (Austin, 1962; Searle, 1969). It is a 'Durkheimian' position as suggested by Kratochwil (1984, p.686; see also Lynch, 1994, p.590). The practices of governing reconstitute and construct the practices of sovereignty in the Arctic. According to Foucault: 'the success of history belongs to those who can seize these rules, replace those who had used them, disguise themselves so as to pervert them, invert their meaning, redirect them against those who had initially imposed them and in controlling this complex mechanism they will make it function so as to overcome the rulers through their own rules' (Foucault, 1977, p.151).

States and Indigenous Peoples

Indigenous Peoples in the Arctic

The countries bordering the Arctic Circle - the United States, Canada, Denmark, Norway, Sweden, Finland and Russia - have indigenous peoples living in the Arctic region. The only exception is Iceland. In North America, the words 'aboriginal' peoples or 'natives' are often used. In Alaska, the indigenous peoples include the Inuit (Eskimos), the Aleuts, the Athabascan Indians and the Southeast Coastal Indians. The Sámi is the only indigenous group in Norway, Sweden and Finland. There is a Sámi population living on the Kola peninsula in Russia. Besides the '26 small peoples' recognized by Russian law, there are many other indigenous peoples in the Russian north and far east (see Dahl, 1993, p.105-107).

The most serious environmental change affecting indigenous peoples and their communities in the Arctic is the constriction of indigenous controlled land. This continues to the present. The loss of land is also a question of losing one's community, way of life and identity: "When land is lost the resource base is diminished, but it also implies an encroachment upon an essential part of the culture itself" (Dahl, 1993, p.125). This development is not a recent phenomenon; the process of losing control over the land has continued for centuries. It is a process characterized by acts of defining and redefining the status of indigenous peoples and their lands, and by granting and denying rights to land through different legislative measures.

For the Sámi, establishing national sovereignty over Sámi land has been seen as a process which also transferred any Sámi rights over land and natural resources to the respective states. The fact that the Sámi never made any treaties concerning their tradition and rights only confirmed the official notion of the Sámi as a being a people without land rights (Brantenberg, 1991; see also Korsmo, 1993; Korpijaakko-Labba, 1989; Henriksen, 1996). The Sámi representative to the AEPS preparatory process emphasized the importance of historical rights and tied historical developments to indigenous identity: 'It is due to the history of colonization of Arctic marginal areas that the indigenous cultures are regarded as users of resources than owners' (Aikio, 1990, p.4).

The Sámi claim to self-determination is the basis for their nation and separate identity as a people. According to the Sámi Council, the Sámi are 'an indigenous population of Sapmi. Our people have inhabited Sapmi from time immemorial, tending land and water with great respect and care'. The claim to self-determination includes the right to decide the use of natural resources. According to the Sámi view as the 'established owners' of the land and water in Sapmi, 'this makes it not only our right but also our duty to protect natural resources for the needs of future generations' (Sámi Programme of the Environment, 1990, p.2). The issue of land rights remains in contention. According to a recent opinion of the Sámi Council: 'The question of Sámi land rights is now the main source of conflict between the nation states and the Sámi peoples. The states involved seem unable to solve this conflict nationally' (Kuokkanen, 1997).

An exception in Scandinavian politics is Greenland. The colonial status of Greenland was abolished in 1953 and, following a referendum, home rule for Greenland was established in 1979. Home rule was defined in territorial terms, which means that electorate of Greenland consists of both resident Inuit and Danes. The home rule authorities have assumed control of health, taxation, industry, transportation, social services, environment and education. The Danish government retains control of defense and foreign affairs. The transition to home rule has been a gradual process and difficulties in achieving autonomy hinged on Greenland remaining dependent on Denmark for economic support (Nuttall, 1992; Dahl, 1993; Larsen, 1992). The process of collectivization under the Soviet rule in Russia led to a disruption of indigenous land use practices and lifestyles in the 1930's (Vakhtin, 1992; Slezkine, 1994). In 1990, Russian president Boris Yeltsin encouraged regional elites to 'take as much sovereignty as you can swallow' and, as Greg Poelzer notes, 'they did'. The process of devolution - transferring power from the central government to local governments - has been accompanied by a policy of economic self-sufficiency. In the past, local aboriginal communities received large subsidies from the state, however, today self-government means also self-financing. There is a crisis of authority. Authority has been transferred to local governments without a corresponding institutionalization of political power and the financial means to govern (Poelzer, 1995, p. 207-212; see also Fondahl, 1995; Osherenko, 1995).

In Alaska, the solution to define the land ownership relationship between the state and indigenous peoples was given in the Alaska Native Claims Act (ANCSA) in 1971. All claims of 'aboriginal title' by the Indians, Aleut and Inuit living in Alaska was extinguished by in exchange of monetary payments and rearranging land ownership in Alaska. Once the land was acquired, natives became the third largest land owners in Alaska (Arnold, 1976, p.160).

Two questions were left open by the ANCSA arrangement: the issues of subsistence and sovereignty. The problem with the arrangement was the hunting and fishing rights, which were considered by the indigenous peoples as essential to their subsistence. The State of Alaska does not recognize any native right to hunt, fish and gather. Congress, however, has enacted special provisions respecting the subsistence activities of the Alaska natives (Flanders, 1989).

ANCSA is a land settlement act and says nothing about indigenous politics and self-government. The tribal sovereignty movement in Alaska wants to revive the traditional and Indian Reorganization Act Councils that had existed since the 1930's. They want to turn their corporation's land over to the tribal government. The advocates of this strategy hope to extend their powers over all aspects of management, including that of fish and game. In contrast, the leaders of native corporations sought solutions that did not involve tribal governments (Morehouse, 1989; Kasayulie, 1992; Fienup-Riordan, 1992).The State of Alaska has argued against the special native rights based on the tribal model of Indian country and 'domestic dependent sovereignty' (see, Price 1982).

The developing relationship between the Inuit and the Government of Canada is of interest for the entire indigenous peoples community. In an agreement made in 1992, the Inuit of the Northwest Territories:

Cede, release and surrender to Her Majesty The Queen in Right of Canada all their aboriginal claims, rights, title and interests, if any, in and to lands and waters anywhere within Canada and adjacent offshore areas within the sovereignty or jurisdiction of Canada (Nunavut Agreement 1992, p.11).

The arrangement accepted in 1992 included a land claim settlement for a region called Nunavut. The remainder of the land was to become Crown land owned by the federal state. The agreement contains monetary compensation. In addition, the Nunavut were to have their own legislature assembly with powers similar to those existing with the Northwestern Territories' Legislative Assembly. Because of the composition of the population in the Nunavut, the model of self-government is based on ethnicity (Nunavut Agreement, 1992, p.235).

The Inuit of Greenland, Alaska and Canada claim to be 'one indivisible people with a common language, culture, environment and concerns'. It is 'only boundaries of certain nations' that separate the Inuit and the 'oneness of its culture, environment and land and the wholeness of the homeland'. The Inuit view is that 'borderlines separating us today were not drawn nor accepted by us. We have never given up our sovereignty' (in Lauritzen, 1980, p.235).

Despite steps in recognizing the economic and political rights of the indigenous peoples of Canada, their rights were not recognized for a long time in the Canadian constitution, although there were several attempts in the negotiations during the 1980's (see, for example, Schwartz, 1986). The 'positive' interpretation of the nature of indigenous rights was stressed in the Royal Commission Report on Aboriginal Peoples in 1996. It proposed calling for the recognition that 'Aboriginal peoples are the original inhabitants and caretakers of this land and have distinctive rights and responsibilities flowing from that status'. The report raises the need for constitutional amendment in the Canadian Constitution Act of 1982, including recognizing 'the inherent right of self-government as an Aboriginal right' (Royal Commission on Aboriginal Peoples, 1996, p.11).

These national differences in relations between states and indigenous peoples suggest that the issue of sovereignty and self-determination cannot be excluded from Arctic environmental cooperation. The claim to self-determination is the basis for the collective identity of the indigenous peoples in the Arctic. The resolution adopted in the Arctic leaders' meeting in 1973 stated that the representatives have recognized their respective identity through these discussions:

We are autonomous peoples, that is, we are an integral part of the very lands and waters we have traditionally used and occupied. Our identity and culture is firmly rooted in these lands and waters. It is this relationship which constitutes the very unique features of our cultural identity in contrast to the cultures of other peoples within each of the countries from which we come (Kleivan, 1992, p.231).

The common Arctic identity of indigenous peoples is based on the idea of 'there is only one Arctic and that we share one future together'. Concern for the health, well-being and ultimate survival of their peoples was voiced at the 1991 Summit of The Arctic Leaders following the first ministerial meeting of the Arctic environmental cooperation. The Arctic leaders required that state governments recognize and satisfy the rights of indigenous peoples to self-government, lands, renewable and non-renewable resources, and to recognize their cultural, social and economic rights (Faegteborg, 1993, p.34).

Protecting Indigenous Peoples

For the Greenlandic representative it was clear that most of the indigenous peoples living within the Arctic have not obtained the position of political or economic self-government. However, having self-government is not enough:

... Even where we do possess such powers internally - like in Greenland - we can only act within our own region and on our sources of possible pollution - and not on the sources, which lie outside our own lands and region (Olsen, 1991, p.2).

Protection of the indigenous peoples and their rights is seen as the responsibility of the Arctic states. The Greenlandic representative noted that:

Only the nation states - and only the nation states of the Arctic in full cooperation - hold the means, and the power to actually do something to counteract the environmental threats to the Arctic environment, thereby helping us to save our very culture (Olsen, 1991, p.2).

The indigenous peoples saw the issue of environmental protection as one of including concerns other than those purely related to the protection of the environment. The representative for the Finnish Sámi pointed out that 'When considering the protection of environment in connection with indigenous peoples it shall be noticed that the issues cannot deal with the natural environment only'. The Sámi representative noted that 'it is important, in this regard, to consider whether the agenda for the proposed ministerial conference would include the issues regarding Arctic native peoples' (Aikio, 1990, p.14). Protecting the environment is also a question of human rights for the Sámi. According to the Sámi, it is possible to regard environmental rights in connection with the land rights and human rights. For the Sámi, 'the problem is that the legal regulation also in the Fennoscandian north is based on the principles of agriculture and not on those of the extensive Arctic culture'. The Sámi view is that Arctic cultures do not enjoy sufficient legal protection and they are often open to free competition where everyone wishes to use the resources of the Arctic nature as effectively as possible. The ultimate conclusion by the states is that

> ... the Arctic nature must be protected from the human influence. This leads to the situation where the Arctic cultures become subject to restrictions although they originally were not hazardous for the survival of the Arctic nature (Aikio, 1990, p. 14).

The Sámi representative emphasized that 'in the vast Arctic circumpolar region a great number of indigenous peoples are living still partly according to their ancient traditions' (Aikio, 1990, p.14). The Sámi representatives point it out that national legislation and regulations can be detrimental to sustainability if the traditional knowledge and indigenous peoples' own relationship with nature are unnecessarily violated (Utsi, 1996, p.2).

The Arctic indigenous peoples have noted that it is not 'fully accurate to imply that countries would fulfil their national and international responsibilities in the Arctic if they ensure the future health and well-being of Arctic ecosystems' (ICC, 1990, p.1).

For indigenous peoples, the question is of a right to development and environment for them. The Inuit consider it a human right to have 'a decisive influence on the development of projects which are planned at, in, or influencing our territory - be it land or sea'. The Inuit saw environmental cooperation in the Arctic from this perspective: 'environmental and human rights have a clear connection in this context' (Lynge A., 1992, p.9). It is an issue of human rights that Inuit traditional knowledge about the environment and living resources is taken into consideration. It is a violation of human rights when the natural resources on which the existence of the Inuit are based are contaminated and by that, threaten their health (see Principles and Elements for a Comprehensive Arctic Policy, 1992).

Northern indigenous peoples are not looking for independence from the states but for the protection of their rights to the environment by the states. For example, the Association of Indigenous Minorities of the North, Siberia and Far East of the Russian Federation is looking for 'a special status' for indigenous peoples including, primarily, 'the existence of a mechanism that does not permit the ignoring of the political and social demands of the Small Peoples' and second, such a mechanism would have 'a legal foundation'. This special status also includes its own structures of self-government for indigenous peoples. The association suggests the establishment of parks and reserves and other forms of protected territories in which indigenous peoples could live 'undisturbed, practice their traditional professional activities and take part in preserving nature' (Sanghi, 1996, p.69-70).

Environmental Rights

The demand of indigenous peoples is to be considered as 'peoples' and claim their right to self-determination. Self-determination is understood as the right to autonomy and self-government in matters relating to their internal and local affairs. For example, in the case of the Nunavut, the agreement include rights of economic activities, land, resource management, and the environment. The agreement recognizes that the Inuit are 'traditional and current users' of wildlife, and their right to harvest wildlife stems from their traditional and current use. The Inuit are also users of certain marine areas, especially the land-fast ice zones. The agreement recognizes that there is a need for a system of wildlife management that complements Inuit harvesting rights and priorities, and recognizes Inuit systems of wildlife management that contribute to the conservation of wildlife and the protection of wildlife habitats. The government, however, retains ultimate responsibility for wildlife management (Nunavut Agreement, 1992, p.26).

The agreement provides the Inuit with a role in international negotiations:

> The Government of Canada shall include Inuit representation in discussions leading to the formulation of government positions in relation to an international agreement relating to Inuit wildlife harvesting rights in the Nunavut Settlement Area, which discussions shall extend beyond those discussions generally available to non-governmental organizations (Nunavut Agreement, 1992, p. 54).

A recent Canadian government report notes that it is 'important that indigenous peoples' representatives have opportunities exert influence at the early stages of policy development, i.e. domestic milieu - which may then lead to the international discussion table' (Report of the House of Commons Standing Committee on Foreign Affairs and International Trade, 1997, p.168). The Canadian government and northern indigenous peoples have clearly arrived to a mutually beneficial arrangement. According to the agreement, the sovereignty of Canada over the waters of the Arctic archipelago is supported by Inuit use and occupancy (Nunavut Agreement, 1992, p.135).

In the Nuuk Declaration in 1993 the states acknowledged Principle 22 of the Rio declaration, which emphasized the role of indigenous peoples in advancing sustainable development (The Nuuk Declaration on Environment and Development in the Arctic 1993, p.4). According to the article 22 of the Rio Declaration on Environment and Development adopted in 1992 in UNCED:

> Indigenous people and their communities, and other local communities, have a vital role in environmental management and development because of their knowledge and traditional practices. States should recognize and duly support their identity, culture, and interests and enable their effective participation in the achievement of sustainable development.

The evolving international standards on indigenous peoples postulate their right 'to the conservation, restoration and protection of the total environment' (Tomasevski, 1995, p.258; 260). The relationship between indigenous peoples, the environment and self-government were recognized by the UN meeting of experts in 1991. Experts concluded that the 'self-government of indigenous peoples is beneficial to the protection of the natural environment and the maintenance of ecological balance which helps to ensure sustainable development' (The Nuuk Conclusions and Recommendations, 1991).

The link between the environment and development was emphasized by the UNCED. According to Principle 3 accepted by the UNCED, 'the right to development must be fulfilled so as to equitably meet developmental and environmental needs of present and future generations' (Rio Declaration on Environment and Development, 1992). The idea of the right to development has been accepted within the context of the United Nations. The UN adopted the idea of the right to development in 1986. According to the Declaration on the Right of Development (1986), development is an inalienable human right. Since 1993, under the Commission on Human Rights, there has been a working group on the right to development. A special rapporteur to the Commission on Human rights notes that 'the human right to development implies as the Declaration on the Right to Development makes clear the full realization of the right of peoples to self-determination and to full sovereignty over all their natural wealth and resources' (E/CN.4/Sub.2/1991/8, p.19).

The right to development is subjective; development has different meanings for different peoples. The emerging right to development has a special significance for indigenous peoples, since the history and impact of 'development' on indigenous peoples, their communities, lands, territories, and way of life has been significant (Sambo, 1992, p.168). In the 1989 International Labour Organization (ILO) Convention concerning Indigenous and Tribal Peoples in Independent Countries, Article 7 states that indigenous peoples

> ... shall have the right to decide their own priorities for the process of development as it affects their lives, beliefs, institutions and spiritual well-being and the lands they occupy or otherwise use and to exercise control, to the extent possible, over their own economic, social and cultural development.

The right of self-determination includes concerns related to the environment. The ILO Convention states that the rights of ownership and possession of the peoples concerned over the lands which they traditionally occupy shall be recognized. Governments are required to take the necessary steps to identify the lands which the peoples concerned traditionally occupy, and to guarantee effective protection of their rights of ownership and possession. The states are also expected to take measures which will 'safeguard the right of the peoples concerned to use lands not exclusively occupied by them, but to which they have traditionally had access for their subsistence and traditional activities' (ILO Convention, 1989, article 14).

The ILO Convention has been ratified by only Norway and Denmark from the Arctic states. The main obstacle to ratification in the other Scandinavian countries is found in the same section that states the rights of ownership and possession of the peoples concerned over the lands which they traditionally occupy shall be recognized (Beach, 1994, p.197).

An earlier human right's instrument, the International Covenant on Civil and Political Rights (1996) can be interpreted in a way that emphasizes the need for states to ensure the rights of indigenous peoples to their lands and natural resources. Article 27 of the International Covenant on Civil and Political Rights (1966) states that 'in countries where there are ethnic or linguistic minorities, persons belonging to such minorities shall not be denied the right to practice their own culture and use their own language'. Whether or not Article 27 of the Covenant also covers territorial rights (the rights to land and water) depends on the understanding of the concept of culture contained within the provision. The question is whether it also covers the material foundation, the economic and physical basis, of the culture of an ethnic minority. If so, the article may include the right to natural resources (Boyle, 1996).

One possible interpretation of this is the idea that each ethnic minority should have the right to demand a real basis that is decisive for just that minority's enjoyment of its culture. This interpretation is based on the view that indigenous peoples are in great need of protection for their traditional use of land and water, because the relationship between the environment of their culture and the exercise of their traditional trades is particularly strong. This view indicates that the use of natural resources and other economic circumstances should be covered to the extent to which this is decisive for the group's maintenance and development of its own culture (Smith, 1991, p.125).

The 1989 ILO Convention recognized the right of indigenous peoples to take participate in the conduct of public affairs and in decisions which affected them directly, such as land rights, social security and health. Article 15 provides for the rights of the peoples concerned to the natural resources on their lands to be specially safeguarded and that these rights shall include the right to participate in the use, management and conservation of those resources. This Article has been interpreted to mean that under the Convention indigenous peoples have a right to control over their lands and they should be granted special political group rights. In particular, those rights should include representative organs and appropriate mechanisms for having 'a real impact' on decisions taken at the national, regional and local levels which effect their identity (see Myntti, 1996, p.24-25).

Garth Netthein points out that international law ought to be sufficient in principle to meet the claims of indigenous peoples: 'it is the implementation of the law which blocks them' (Nettheim, 1988, p.120). The UN experts affirm this view:

> Although there is still a need to develop rules in order to lay a legal foundation for this right and improve the machinery for its protection, the fact remains that there are enough frameworks for action in existence at the present time for it to be effectively implemented (E/CN.4/Sub.2/1991/8, p.29).

One reason for this is, according to the UN rapporteur, that non-participation in decision-making at the international and national level has been and remains at the root of imposed development choices or strategies that have done grave harm to the environment. However, the principles of national sovereignty, territorial integrity and the political unity of independent states are often preferred at the expense of the principle to the self-government of indigenous peoples in different national practices (Myntti, 1995, p.137-138).

Practices of Power

'Indigenous'

Fredrik Barth's focuses on boundaries used to maintain and construct ethnicity. Ethnic groups are categories of ascription and identification by the actors themselves and thus, have the characteristic of organizing interaction between people. Barth stresses the different processes in generating and maintaining ethnic groups by constituting ethnic boundaries. He focuses on internal maintenance of boundaries, but notes that ethnic groups are connected to each by interdependent relations which may be differ from co-residence and competition to reciprocity (Barth, 1969, p.16-17).

Making boundaries and defining identities is an act of power. From the Foucaultian standpoint, 'indigenous' in political terms is a statement of power. The principle of indigenousness may be defined essentially in political terms that acknowledge the special status of the original occupants of a territory and aims at resorting rights and entitlement that flow from recognition of this special unique relationship with the state. It provides the 'theory' as well as the practices for redefining indigenous-state relations; self-determination and self-government is the practical expression of this theory (Fleras and Elliott, 1992, p.30).

Jens Dahl makes a difference between the rights of indigenous peoples in their relations with the states and the relationship that they have with the environment. He stresses that cultural rights should be kept separate from rights based on state legal systems or specified rights to natural resources. Self-determination, Dahl (1996, p.17) claims, is 'first of all and act of cultural identity and secondly a right to be claimed within national and international law'. Dahl makes a clear difference between cultural identity and the legislative identity of indigenous peoples.

Foucault's idea of identity is the opposite of Dahl's understanding. The relationship between states and individuals is defined in terms of 'rights': The focus is on how indigenous identities are defined by legislative measures and by granting rights. In Foucault's terms, 'what enables the identity of a community to be defined is political practices of subjection' (Simons, 1995, p.53).

The individual to which power has been constituted is its vehicle. As a human subject is put into a position of an object he or she is placed in a complex network of power relations. The individual is not the vis-a-vis of power; it is one of its prime effects. The state acts on the individual but in the process a new, general type of individual is produced. Foucault does not deny the negative repressive force of the state. He insists that power relations are more complex. The state segregates, labels, creates and oppresses (Foucault, 1980a, p.98).

The techniques of self also include resistance. The individuals can be seen as 'self-determining agents who are capable of challenging and resisting the structures of domination in modern society' (McNay, 1992, p.4). However, 'there are no good subjects of resistance,' therefore nobody is innocent, original or outside the grid of power (Foucault, 1980b, p.257).

This view does not look for origins as an epistemologically problematic quest for ahistorical and asocial essences. The search for the origin of a particular historical phenomenon implicitly posits some form of original identity prior to the flux and movement of history. The Foucaultian approach emphasizes a nonessentialist definition of identities; identities are made by political decisions and through legislative documents (Foucault, 1980a, p.98).

For governmentality the word indigenousness does not refer to 'origins,' since all human beings are indigenous in some way, but here, it is a reference to groups that occupy a determined position in society because of specific historical developments. The word 'indigenous' refers to a set of relations between states and population groups as a historical continuity from the occupation of the territories inhabited by these groups to the demands for their specific rights because of injustices suffered and for their special quality as being indigenous - claiming to be the original occupants of the region (Stavenhagen, 1994, p. 14; 16-17).

The word 'indigenous' refers to both descendants of the original inhabitants of a territory which has been overcome by conquest and to peoples who have characteristics of a national minority including a common language, religion, culture and other identifying characteristics, and a relationship to a particular territory but are subjugated by a dominant culture and society (Assies, 1994, p.49). A minority is 'a group numerically inferior to the rest of the population of a state, in a nondominant position, whose members - being nationals of the state - possess ethnic, religious or linguistic characteristics differing from those of the rest of the population and show, if only implicitly, a sense of solidarity, directed towards preserving their culture, traditions, religion or language' according to Article 27 of the International Covenant on Civil and Political Rights (1966). Although there is an overlap in the distinction between indigenous peoples and minorities, 'indigenous' refers to peoples who have been affected by the past 500 years of colonialism. Most indigenous peoples satisfy definitions of 'minority' but many minorities are not indigenous (Thornberry 1991, p.331; see also Sanders, 1993).

Indigenous peoples claim to have a special relationship to the environment and to the land where they have lived for millennia. Indigenous peoples are people 'who are regarded as indigenous on account of their descent from the populations which inhabited the country, or a geographical region to which the country belongs, at the time of conquest or colonization or the establishment of present state boundaries and who, irrespective of their legal status, retain some or all of their own social, economic, cultural and political institutions' (see ILO Convention, 1989, article 1).

The perception of indigenous peoples as a 'problem' to be solved constitutes the central motif in the evolution of statehood in the Arctic. Strategies for the solution of the problem at the national level have changed from a commitment to assimilation by segregation, wardship and protection, through to the era of integration and formal equality in the post second world war era and the arrangements for limited autonomy in 1970's and 1980's. Developments in the circumpolar region have common features, although the phases and modes of policies are different. The tradition of how states treat the indigenous peoples of the Arctic has been based on using different definitions and policies for different groups of indigenous peoples in national legislation. This tradition is continued in the international practices of cooperation within the circumpolar region.

The contribution of the Foucaultian approach to the study of power is to show how relations of 'agency' and 'structure' have been constituted discursively: how agency is denied to some things and not to others, and how structures could be said to have determined some things and not others. Foucault's theory is extremely constitutive but neither agency nor structure is given primacy (Clegg 1989, p.158).

This view questions the macro-level analysis of power relations suggested by international relations theorists and suggests that power should be studied as an event, in a different space-time form to its emergence, at the microlevel. Typical to the traditional formulation of power is to reduce it to those who have the power in terms of material or intellectual resources or to the institutions of power. Power, in the Foucaultian view, is not the possession of a certain amount of material capability but is exercised by drawing upon material and authoritative resources (Foucault, 1980a, p.95-96). This view challenges the traditional view of international relations theory on power according to which power was which as seen, shown and manifested. This approach sees international politics as social rather than strictly material (Wendt, 1987, p.360; Dunne, 1995a, p.371-372).

Power is a name given to the complex strategic situation in a society. Power is an 'intrinsic element of interdependence' (Milner, 1991, p. 83; Rood, 1989, p.73; Keohane and Nye, p. 1977, 11). Intersubjectivity and considerations of power are not separate; power does not exit independently of human relations. Power does not depend on consent; the relationship of power can be the result of a prior or permanent consent but it is not by nature the manifestation of a consensus. Power has to be understood as a process and a continuous struggle. Power is productive; it is a network of relations structuring a phenomenon, defining a space of interaction or a field of possibilities. It is not only force that restrains and constrains actors. Rather, it is a force that also produces something (Foucault, 1982a, p.220-224).

Power is nonsubjective. The question is not who has power, it is how is it that subjects are constituted (Foucault, 1980a, p.97). Power exists 'between' people: 'the 'other' (the one over whom power is exercised) is thoroughly recognized in the end as a person who acts; and that, faced with a relationship of power, a whole field of responses, reactions, results and possible interventions may open up'. Power resides in the inter-world; '... it is always a way of acting upon an acting subject or acting subjects by virtue of their acting or being capable of action. A set of actions upon other actions' (Foucault, 1982a, p.220).

Indigenous Peoples' Participation

At the first consultative meeting in Rovaniemi 1989 the participants were officials either from different ministries of the environment or of foreign affairs. Some national delegations included researchers as delegates. The issue of the involvement of indigenous peoples was discussed in the meeting. There was no one solution to this issue, in that: 'different situations in the Arctic countries may call for different ways of achieving this'. However, it was considered that 'indigenous peoples should be involved in future work since they bear the burdens of environmental degradation directly' (Consultative Meeting on the Protection of the Arctic Environment, 1989, p.6).

A further problem has been in defining which groups should be included into the process as representatives of indigenous peoples. It has not always been clear who the indigenous peoples are in the Arctic. As the Danish representative noted:

> I think in a way we are all indigenous but in different places. We all want to influence our own fate and protect our homelands just as the indigenous peoples of the North want. These endeavors are our endeavors (Haarder, 1991, p.4).

This idea of doubting (or questioning) the demand of indigenous groups, as I interpret the previous statement, has changed considerably within the AEPS experience. Recognizing the role of indigenous peoples has produced a new word in the AEPS language: IPO's (Indigenous Peoples Organizations). The recognition of IPO's as actors in the process is an acknowledgment of the fact that these organizations have played a major role in developing the AEPS (Petersen, 1996).

The Danish contribution to indigenous peoples' participation was important. Denmark stressed that it shared 'Arctic responsibility' in close cooperation with Greenland. It is therefore 'impossible to talk about Danish Arctic policies without simultaneously looking at what is going on in Greenland'. The Danish representative emphasized that: 'To a considerable extent Greenlandic Home Rule policies simply are Danish policies' (Statement by the Danish Delegation, 1989, p.2). The Sámi Council representative stresses that the efforts and pressure by Sweden on other Scandinavian countries to involve indigenous peoples as observers was important (Halonen, 1996a).

It was the Canadian representative who suggested that 'sufficient attention is given to the interest, roles and involvement of northern regional governments indigenous peoples, industry and non-governmental organizations' (Statement by the Canadian Delegation, 1989, p.3). The idea became real at the Yellowknife preparatory meeting where some Inuit and Sámi representatives were present. The Inuit representative considered as important the 'direct participation of indigenous peoples at all stages of an Arctic sustainable and equitable development strategy, so that indigenous perspectives, values and practices can be fully accommodated' (ICC, 1990, p.2).

The Inuit representative suggested, 'an effective cooperative process that would ensure the direct involvement of indigenous peoples in the Finnish initiative'. Different steps were suggested to involve indigenous peoples in the process: 1) distributing information about the process itself, 2) preparing a detailed proposal on indigenous participation, 3) giving financial support for indigenous participation, and 4) organizing a preparatory meeting of indigenous peoples. The wish of the indigenous peoples' organizations was for indigenous peoples' relations with state governments to be based on 'principles of cooperation and respect, rather than on unilateral state action' (Simon, 1990, p.3).

Three Arctic indigenous people's organizations have been 'permanent observers' within the process since 1991. The Inuit Cirucmpolar Conference (ICC) represents the Inuit way of life in Alaska, Canada and Greenland. It was established in Barrow, Alaska 1977. The Inuit residing in the Russian republics were accepted as full members of the ICC at the 1992 meeting. The ICC concentrates on international relations and issues concerning the Inuit, in contrast to regional Inuit organizations (Creery, 1994, p.143-144; see also Lauritzen, 1983).

The Association of 26 Small Peoples of the Soviet North was founded in Moscow in March 1990 (later known as the Association of Indigenous Minorities of the North, Siberia and Far of the Russian Federation). The association brings together the representatives of the different regional organizations established at the end of 1980's (Dallman, 1994, p.58; see also Alia, 1991). The association was not present at the Yellowknife meeting. However, the Soviet representative mentioned the association at the meeting. The role of the association was described as 'to take care of the interests of the indigenous population at all levels of government' (Statement from the Soviet Delegation, 1990, p.4). There was readiness by the Soviets to accept the participation of the association in the process since, according to the representative, 'appropriate policies cannot be worked out without the participation of indigenous peoples of the North' (Statement from the Soviet Delegation, 1990, p.9).

The Sámi Council represents the Sámi way of life in the Scandinavian countries of Finland, Sweden, Russia and Norway. The Nordic Sámi Council was established in 1956. The Sámi on the Kola peninsula were organized in 1989, when it became possible for them to organize politically (Sergejeva, 1995). The participation of the Sámi on the Kola Peninsula since 1992 has led to the renaming of the organization as The Sámi Council. The Council is responsible for promoting commonly held Sámi views for general knowledge and public discussion (Minde, 1995; Seurujärvi-Kari, 1994; Sillanpää, 1994).

The indigenous peoples considered the Rovaniemi ministerial conference in 1991 'historical,' since it was the first time that indigenous peoples of the area participated in the preparatory process of making of a joint international declaration on Arctic environmental conservation (Alkuperäiskansojen yhteistyö tiivistynyt arktisessa ympäristönsuojelussa, 1991). According to Aggaluk Lynge, the head of ICC delegation, 'this was a great step for us' (Alkuperäiskansat mukana arktisten alueiden ympäristökonferenssissa, 1991). The AEPS was characterized as 'a long-awaited step in a mutually beneficial direction' by the representative of the ICC. Lynge (1991, p.2) considered the participation of indigenous peoples as important since 'where the indigenous peoples have been displaced or otherwise lost control of their lands, we see that environmental destruction quickly follows'.

The Sámi Council representative described the contribution of indigenous peoples thus 'We the Sámi are able to agree and cooperate with the Arctic governments in order to combat pollution'. States are, however, asked to recognize that 'the restrictions which may be put through in future legislation would give us the opportunity to live in the Circumpolar Arctic utilizing the natural resources, our land, water and territories with our traditional technology' (Halonen, 1991, p.2).

The representative of the Greenlandic Home Rule from the Danish delegation noted that the involvement of the indigenous peoples in the process has shown 'an absolute positive attitude towards the Arctic indigenous peoples'. This positive attitude was also shown 'by incorporating our organizations in the preparatory work and by securing our continuing participation in the future work' (Olsen, 1991, p.2).

Indigenous Peoples as Permanent Participants

During the negotiations to establish the Council from June 1995 to August 1996, the theme of participation by indigenous peoples' was one of the major items on the agenda. The source of the 'problem' in the negotiations was the United States who, in the meeting in Washington, D.C., opened the discussion by bringing a suggestion to add two new groups from Alaska (the Athabascans and Aleuts) to the category of 'permanent participants' in the Arctic Council. The discussion evolved around the issue of defining the access requirements for the status of a permanent participant, their number and involvement in the work of the Council. The progress made in previous negotiations was lost and the 'Washington meeting was a disappointment' according to a Finnish representative (Puurunen, 1996).

For the U.S. it seemed most important that 'they [indigenous peoples] are there' to participate. According to the U.S. SAAO (Senior Arctic Affairs' Official), the 'U.S. encourages indigenous participation into working groups'. Senseney described the negotiations on the issue as 'a long and difficult discussion [on] which groups are represented'. The issue was difficult, particularly concerning representation from Alaska. The term 'Alaska Natives' includes the Yupik, Inupiat, Aleut, Athabascan, Tlingit, Haida and Tsimshian peoples. They speak several different languages. Over half of them live in nonurban areas. Alaska Natives can be found in a variety of occupations and ways of life, from subsistence hunting and fishing to state and corporate offices. There are over 200 tribes, twelve regional native corporations and 200 village corporations in Alaska. Some Natives, though, feel that they are not truly represented by any of these organizations (Senseney, 1996).

The aim of the Aleuts and Athabascans was 'a representation on a par with three major indigenous groups'. The problem mainly concerned the way the organization of indigenous communities was instituted on the North American side of the Arctic. The ICC does not represent all the indigenous peoples and their local organizations in North America, and only represents part of the indigenous population of Alaska. As a result, the Inuits do not have a mechanism in place for consultation with, or the provision of, constituent services with the other groups. Only the Inuits and the Athabascan reside within the geographic Arctic in Alaska but the U.S. changed its definition of the Arctic for the AEPS, by enlarging it to include the Aleutian chain in 1995. This resulted in the Aleuts living on the 120 islands of the chain becoming 'Arctic' residents for AEPS purposes (Senseney, 1996).

The indigenous groups were looking for 'closer involvement with the activities of the Arctic Council'. Flore Lekanof from the Aleutian/Pribilof Islands Association in Alaska stated the U.S. position that the Aleuts and Athabascans should be recognized by allowing each of their organizations to have 'Permanent Participant' status in their own right. Lekanof expressed concerns over the language used in the proposal especially the meaning of 'non-governmental'. In his view it would disqualify all tribal governments from applying for permanent participant status. In February 1996, the Aleutian/Pribilof Island Association described the present situation as 'hardly representative' since they 'don't have a voice'. The idea of having joint representation with other groups (the Dene and Athabascans, for example) did not get them very excited: The 'Aleuts is not a land-based culture as the Athabascans are, therefore, the interests of the groups are different and joint representation could be difficult' (Lekanof, 1996).

For Randy Mayo, who represented the Council of Athabascan Tribal Governments, it was clear that 'ICC does not want any other indigenous peoples' organizations'. He stated that the role of new indigenous peoples' organizations were relevant to the AEPS and the Arctic Council: 'We wanted to get included because of this unsustainable resource use that is having an impact to us'. The representative of the Athabascans was not happy to be a member of the U.S. delegation since 'they do not agree with the U.S. agenda'. The Athabascans want to 'stand on their own' (Mayo, 1996).

The Canadian Arctic Ambassador saw that be 'the permanent participant category is meant exclusively for indigenous organizations'. A provision was kept open for 'additional membership'. The problem is that 'national groups do not have members representing indigenous peoples in their delegations'. A solution had to be found to the question 'how to bring in other groups' into the negotiations. Indigenous involvement was a delicate issue but it was 'indigenous peoples that have to find a mechanism of their own to deal with the issue' (Simon, 1996).

From the indigenous point of view, 'to allow everybody there would result in a disaster' (Huntington, 1995). The fear was that by allowing other participants, the 'voices of indigenous peoples are diluted'. Chester Reimer from the ICC stated that the 'U.S. position has been unclear as well as other states have been also unclear'. There was some frustration among the ICC members since, as Reimer described, 'things were agreed but they were not' (Reimer, 1995). Several people who represented other countries saw the U.S. initiative as obstructive to the entire process of establishing the Arctic Council.

For Margie Gibson from the Arctic Network, an organization that brings together both indigenous groups and environmentalists, expanding indigenous participation is 'democratizing the process,' and not a move by the U.S. negotiators to obstruct the process itself (Gibson, 1995). The explanation for this is in the internal situation in Alaska and the United States; there are 'internal reasons in the U.S. for taking up the participation issue'. Gibson referred to the 'trust relationship' between indigenous peoples and the federal government in the U.S. This relationship obligates the U.S. State Department to ensure that indigenous peoples have an adequate representation on the council (Gibson, 1996).

On the Canadian side, it was noted that the 'Dene and Athabascans do not have as good access to the initiative as the Inuit' (Fenge, 1995). Canada's other indigenous groups were asked by Canada to postpone any request for 'permanent participant' status until a later date. The Canadian position was that a political decision concerning indigenous representation was made in Rovaniemi in 1991 (Mills, 1995). The official Canadian view was that the current three groups represent the major part of the indigenous population in the Arctic. Canada believed that other indigenous peoples' groups should make their input as members of national delegations (Whitby, 1996).

Russian negotiators echoed the concern of Canada in the Arctic Council negotiations, stating that opening the category of permanent participants would mean that 40 groups could apply for accreditation. Several indigenous northern groups in Russia were not currently represented: the Chukchi, the Chuvan, and Yukaghir, the Yakut, the Evenk and the Nenets are not then in the AEPS. It was also unclear for the Russian representatives on just how many additional groups from the Russian North might apply for permanent participant status (Senseney, 1996).

Finally, the issue was left for the indigenous peoples to be solved. The main points in the ICC proposal for a solution were: 1) there were a finite number of indigenous organizations that could be accommodated, 2) there were a potentially large number of interested Arctic peoples' organizations, and 3) certain Arctic indigenous peoples' organizations should have standing as permanent participants (Simon, 1996; Meeting of the Working Group on Permanent Participants, 1995).

In practice, permanent participation:

is equally open to other Arctic organizations of indigenous peoples with a majority of Arctic indigenous constituency representing a) a single indigenous people resident in more than one Arctic state; or b) more than one Arctic indigenous peoples resident in a single Arctic state (Declaration on the Establishment of the Arctic Council, 1996).

The number of 'Permanent Participants' should, at any time, be less that the number of member states. The number of indigenous peoples organizations cannot be more than the number of states whether there are joint seats or separate representation (Declaration on the Establishment of the Arctic Council, 1996).

Among the indigenous peoples, the arrangement of permanent participants was met with satisfaction. According to the Sámi representative:

The status that indigenous peoples and Sámi Council has been given is the highest than ever before for any indigenous peoples' organization in the world ... We can sit at the same table as the ministers with the highest officials when decisions are discussed and made. We have the right to make a statement and we can make whatever suggestions we want to. We only miss the right to make decisions, that is we are not part of the consensus (Halonen, 1996b, p.4-5).

For Arctic governmentality, defining power relations in a traditional way through focusing on the states' relations would have given a rather simplified and narrow understanding of the relations among the actors. Seeing power as microsocial, strategic, productive and as a restrictive phenomenon gives a more diverse picture of the relations among the Arctic actors in the cooperation. Overall, governing is the effective use of power; the states asserted their power and influence through the measures chosen. The reservations of individual countries towards the issue of state sovereignty were respected. For Foucault, governing is the right manner of arranging things so as to lead not to some form of the common good, as suggested by the governance theorists, but to an end which is 'convenient' for each thing that is to be governed (Foucault, 1991b, p.95).

Governing of the environmental issues followed the logic of convenience; there were many reasons for the states for not taking regional actions and for emphasizing national measures and joint action in international fora. Not much could be done, at least according to the states, to the concrete problems. Convenience is seen in the commitments made in the cooperation. States had 'good' reasons for that: no real gaps in the existing legal framework, avoiding overlapping work, and the need to act actively in other international relevant fora in order to advance the protection of the Arctic environment. Convenient governing of the Arctic consists of the guidelines, strategies and applications for the moment. They at the minimum created a sense of actions by the state to deal with the issues.

The 'things' to be governed are not only the territory, but 'a complex composed of men and things, men in their relations, their links, their actions with other things which are wealth, resources, territory etc' (Foucault, 1991b, p.93). This complex includes the relations between the state and indigenous peoples in Arctic environmental cooperation. Interpreting these claims and international practices in the Foucaultian sense focuses the essential function of the discourse and techniques of rights' in defining the relationship between states and indigenous peoples is to 'efface domination' intrinsic to power as the legitimate right of sovereignty, and as the legal obligation to obey it (Foucault, 1980a, p.95).

Defining these rights is the process of reconstituting the relationship between states and indigenous peoples. The practices of power can be also detected in the efforts to redefine the state-indigenous peoples' relationships in the Arctic. The relations between states and indigenous peoples are things to be 'governed'. More important than finding solutions to the problems at hand was to secure existing power relations both inside and outside individual Arctic states. Indigenous peoples were defined first 'observers' and then 'partici-pants' in the Arctic cooperation while the states were the main actors - 'the members'- of the cooperation.

3 Discourse on Knowledge

The Need for Knowledge

The Problem of Uncertainty

Information about the state of the Arctic environment was 'incomplete' and even 'fragmentary' in 1989. Iceland noted that the ability completely to understand pollution issues in the Arctic was restricted by the lack of a comprehensive scientific data base and coordinated monitoring program on the state of Arctic ecosystems (Gudnason, 1991, p.4). Uncertainty about the problem or problems in the Arctic was the basis of the concern:

> What has been especially worrying is the fact, that most of the environmental threats to the Arctic seem to be coming from the outside world - and that there has been no clear picture of the sources and transport patters of the pollutants accumulating in the Arctic (Olsen, 1991, p.1).

States agreed to prepare individual reports on the different concerns of pollution.

The report on organochlorines, which are a large class of manmade chemicals, stated that 'virtually all organochlorines detected at southern latitudes have also been found in the Arctic'. These contaminants were considered 'hazardous' due to their high persistence in the environment, their potential for bioaccumulation and biomagnification, their relatively high toxicity. The amounts released into the environment were considered large. These contaminants reach the Arctic via long-range transport from the more industrialized centers. It was evaluated that concentrations of organochlorines are generally lower in the Arctic than in heavily polluted areas. There were some exceptions; some of these pollutants were often detected in the Arctic at concentrations similar to those in the 'source regions'. Despite the lack of information, however, the report concluded that 'species at the top of the food chain are considered most at risk'. Due to the slow turnover process in the Arctic the rate of a contaminant breakdown may be considerably slower than in temperate ecosystems. Therefore, pollutants may present 'a prolonged threat' to the Arctic environment. Local consumers of wildlife foods are threatened (Jensen,

1991, p.337; 369; 374).

Pollution resulting from heavy metals was considered 'not critical at the present time'. The main input of heavy metals is by their transport with air flows from the industrial and densely populated areas of Eurasia and North America. Heavy metal levels decrease from south to north, but smaller values are found in the Greenland-Scandinavian area. The development of an economic infrastructure and industrialization in the Arctic has resulted in 'an increased anthropogenic load' of pollutants on the ecosystems of the region. An input of heavy metals from local to freshwater sources and an accumulation of heavy metals in tundra soils, snow cover and glaciers has been observed. The threat to human beings was noted; the use of local species of fish, marine mammals, terrestrial mammals and migratory birds with enhanced levels of heavy metals in their tissues presented 'a direct threat to the health of the population' and required more detailed and extensive studies on the harmful effects of heavy metals (Melnikov, 1991, p.112-113).

The reports on acidification and noise were more directly concerned with their impact on the environment. The main anxiety is over acidification caused by transboundary pollution. The available information available was 'incomplete'. It was estimated that anthropogenic influence is restricted but may cause considerable, and possibly irreparable, 'local and regional damage'. Global impact, however, had not been studied. Available information indicated the rate of acidification was be divided into three main categories. First there was large scale acidification, which mainly involved the north-boreal zone. The Taymur Peninsula near Norilsk in Russia is 'an extensively polluted and acidified region in the tundra zone'. Second, acidification is evolving into 'a prominent environmental problem' in Fennoscandia, especially in Finnish Lapland and Norwegian Finnmark, the eastern part of Canada and in the northwestern parts of Russia. Third, there is a south-north or a west-east direction in acidifying influences, reaching sparsely populated and nonindustrialised regions, but those regions also have the lowest depositions. The role of constantly increased emissions produced north of the Arctic Circle is 'unknown' (Nenonen, 1991, p.63-64).

The report on noise noted that the waters of the Arctic region make a unique noise environment, due mainly to the presence of ice. During the periods when ice cracking and wind noise are absent, the areas covered by shore-fast ice are among 'the quietest underwater environments'. Manmade noise is created by ship traffic, offshore activities, airplanes, and small boats. Most important is the traffic north of Russia, where most of the polar icebreakers operate. Other shipping activities include fishing and military activities. The exploration for oil and gas creates underwater noise. These activities create noise levels which may disturb marine mammals, or mask the natural sounds that are important to those mammals. Noise from human activities may sometimes cause short-term behavioral reactions and temporary displacement of various marine mammals. The report indicates that there is much evidence to show that most types of disturbance do not cause mortality. Marine mammals seem able to habituate, or at least tolerate, increased noise levels. The scarcity of direct evidence of serious consequences from noise and disturbance, however, does not necessarily mean that marine mammals are not stressed or affected in another way (Davis et al., 1991, p.157; 246-247).

Future-related threats to the Arctic were oil pollution and radioactive contamination. According to a study on oil pollution, there is 'a lack of good data' on the trends in oil inputs into the Arctic Ocean. The report concludes that 'acute oil spills show large variations from one year to another, while the data available on long-range transport pollutants indicate fairly stable annual inputs'. Oil pollution is caused by discharges of oil-based drill cuttings, ballast water from storage tanks, drainage, and minor spill or leaks of oil during normal operation. Most oil spills, small and large, are statistically linked to the transport of oil by ship (Futsaeter et al., 1991, p.274-276).

The problem of radioactivity was, according to the report, considered 'well studied' in the northern regions. These studies have included direct measurements of radiation and the collection of air, precipitation, water, soil, sediment, fauna and flora samples. They have shown that 'some plants and animals accumulate appreciable amounts of radionuclides as a result of the nutrient poor environment' in the Arctic. Populations that use caribou or reindeer as their main sources of meat also accumulate high amounts of radionuclides, particularly caesium 137. The current levels of radionuclides in the Arctic and its seas do not pose 'a widespread' threat to human health or the environment. The present concentration level in the Arctic reflects contribution from at least four different sources: global fallout, transport from European reprocessing plants, transport by rivers and fallout from Chernobyl. The dominant source of anthropogenically produced radionuclides observed in the Arctic is the fallout from previous atmospheric nuclear weapons testing. Available information also reveals that nuclear material has been disposed of into the Arctic seas (Paakkola, 1991, p.387; 396-398).

Arctic Monitoring and Assessment Program

The states decided to establish a separate program. The idea of monitoring and the assessment of inputs, outputs and their effects in a 'comprehensive and coordinated manner' was proposed during the preparatory process. It was suggested that this might lead to a follow-up program to monitor the state of the Arctic sea, its coasts and fauna and flora, according to the Finnish proposal (Statement by the Finnish Delegation, 1989, p.1). The participants recognized that 'there was a need for coordinated monitoring of important environmental characteristics and pollutants throughout the Arctic' (Protecting the Arctic Environment, 1990, p.8).

The mandate of the AMAP is based on 'an urgent need to improve cooperation among local and regional efforts and global programs in order to better and more complete the understanding of the environmental situation in the Arctic' (AEPS, 1991, p.31). The program was modeled on the UN Economic Commission of Europe (ECE) program on transboundary air pollutants: 'A coordinated monitoring program for air pollutants... would be of interest not only to the Arctic region' but also to ECE's activities to monitor long-range transboundary pollution, 'because of its links with transatlantic transport of air pollution' (Statement of ECE, 1989, p.1-2).

The image of the Arctic and the threats to the environment did not noticeably change as a result of the AMAP publication of 'Arctic Pollution Issues: A State of the Arctic Environment Report' in 1997 (AMAP 1997), just before the AEPS ministerial meeting in Alta, Norway. The Arctic is characterized by 'a harsh climate with extreme variation in light and temperature, short summers, extensive snow and ice cover in winter and large areas of permafrost'. The plants and animals of the region are adapted to these extreme conditions but adaptation is sometimes making them more sensitive to human activities (AMAP, 1997, p.vii).

The AMAP report is a 'consensus report'. It is a compilation of the work of hundreds of scientists (Stone, 1997). The AMAP report is described as, 'both a process and a product'. That is, the chapters of the report are intended to summarize and analyze the contemporary state of knowledge on the sources, levels, distributions, trends, fate and effects of contaminants and other anthropogenic influences on the environment and human health. In addition, the AMAP was asked to assess the relative magnitude of damage and threats to the environment and human health based on existing information. The assessment also included the recommendation of the AMAP for future actions to reduce damages and threats and identify deficiencies and gaps in information and data. The AMAP assessment aims to provide a 'baseline for understanding the status of contaminants in the Arctic'. It is not intended to be an assessment of the environmental impact of contaminants nor a risk assessment for developing risk management policies (AMAP, 1997, p.3).

The AMAP report stresses the global threats to the Arctic environment and the interdependent relations between the Arctic and rest of the world.

Contaminants enter the Arctic from 'outside'. Here, 'outside' refers to sources in Europe and North America. Outside the Arctic, there are a number of sources for persistent organic contaminants and heavy metals. Industrial activities, energy production and transport in areas far from the region produces contaminants with low but widespread levels which arrive in the Arctic. The Arctic is described as a region where many pathways - atmospheric, riverine and marine - transport contaminants into, and within, the Arctic. The Arctic is, therefore, 'a potential contaminant storage reservoir'. Strong south-to-north air-flows, particularly over west Eurasia in winter, transport contaminants from lower latitudes. Arctic rivers are also significant pathways for contaminant transport to the Arctic. Finally, ocean waters are 'a major storage reservoir' and transport medium for, for example, water soluble persistent organic contaminants. The report considers that sea ice may be important in transporting contaminants. The sea 'is the final resting place for most contaminants' (AMAP, 1997, p.vii-viii).

With regard to the 'global' threats of climate change and ozone depletion, the report emphasizes climate change as being of 'immediate interest to the Arctic'. Climate change is 'likely to be more pronounced in the Arctic than in other areas of the world' (AMAP, 1997, p. 161). In addition, feedback mechanisms that can enhance the warming caused by greenhouse gases make the Arctic important for understanding global climate change. In adidition, ozone depletion has been 'more severe in the polar regions than elsewhere in the world'. Reflective snow-cover in the Arctic makes the effects of changed radiation especially pronounced in the Arctic. The effects of the depletion of the ozone layer on humans are well known.(AMAP, 1997, p.180).

According to the report, observations from snow cover and permafrost cores suggest that some warming is already taking place in the Arctic. Temperature records show warming in some areas but cooling in others. The effects of global climate change on Arctic temperatures and precipitation patterns are, according to the AMAP report, 'very difficult to predict' but most studies suggest that the Arctic as a whole will warm more than the global mean. Warming in the troposphere is expected to be followed by a decrease in temperatures in the stratosphere. Temperature records for the lower stratosphere in the Arctic reveal 'dramatic changes'. They are described as the 'steepest decline on the entire planet' (AMAP, 1997, p.161).

The Arctic Ocean has also warmed slightly over the past decade. There is, however, a cooling trend in the Nordic seas. The ice sheets of Greenland hold great amounts of water. Measuring their volumes is very difficult but 'so far they do not seem to be shrinking and might even be growing slightly'. This does not necessarily, according to the report, contradict indications of a warmer climate, since increased snowfall adding more mass to the top of a glacier may compensate for any extra melting and calving of icebergs. Finally, as one aspect of climate change, it is noted that precipitation has increased in high latitudes over the past 40 years (AMAP, 1997, p.160-161).

Some problems are 'regional' in character, according to the AMAP. Some known sources of persistent organic pollutants and two-thirds of the heavy metals in air in the high Arctic originates from industrial activities on the Kola peninsula, the Norilsk industrial complex, the Urals and the Pechora basin. Industrial activities in northwestern Russia, including the Kola peninsula, and at Norilsk are 'the dominant sources of sulfur' north of 60 degrees latitude. The 'major anthropocentric source of hydrocarbon contamination' in the Arctic is oil and gas development. Excepting catastrophic releases of oil, concentrations of hydrocarbons associated with anthropogenic inputs have been relatively slow. Contaminants are 'widely but not uniformly distributed' around the Arctic. Geographical variations in levels result from point sources of contamination, which in turn result in 'high local pollution concentrations,' and from environmental convergence mechanisms (AMAP, 1997, p.viii-ix).

Some problems, by their nature, are 'local'. For example, the environmental threats to the Arctic associated with oil and gas development, production, and transport are primarily local and/or regional and on a circumpolar scale. Local sources of radionuclides, such as dumped nuclear waste, nuclear storage sites, accidents and past explosions have led to local radioactive contamination. Several radioactive sources exist in north western Russia. They are 'a potential threat' for the release of considerable quantities of radionuclides (AMAP, 1997, p.viii).

For the standard regime theorists, this information gives actors a sufficient consensus to serve as a guide to public policy designed to achieve some social goals (Haas, 1980, p.367-368). Regimes are assumed to include essentially objective information or standards for behavior which the parties have to live up to (Haas, Williams and Babai, 1977, p.12). The information has an instrumental function in international environmental politics since, for example, Haas (1975, p.858-859) argues that 'scientific knowledge will create a consensual basis for the recognition of new cause-effect links which had not been recognized before'. These beliefs are accepted irrespective of their absolute or final truth. Shared perceptions, beliefs and understandings of causal mechanisms of problems among the relevant parties are important in the regime formation (Osherenko and Young, 1993, p.20).

The mainstream approach to knowledge in international relations theory is in viewing it as 'consensual'. Consensual knowledge is understood as a body of beliefs that are widely accepted by the relevant actors. Consensual knowledge is 'generally accepted understanding about cause and effect linkages about any set of phenomena considered important by society, provided only that the finality of the accepted chain of causation is subject to continuous testing and examination through adversary procedures' (Haas, 1990, p.21).

Creating the Consensus

The pollution data available to the AMAP from the Arctic region is mainly based on national programs carried out within limited areas. The AMAP Project Directory includes approximately five-hundred monitoring and research projects and programs of relevance in the Arctic area, of which approximately three-hundred comprise the AMAP national implementation plans. The information in the AMAP reports is also based on existing literature, since some AMAP projects have only recently been started and have yet to produce adequate information for assessment. The initiation of new programs was restricted (Reiersen, 1996).

The monitoring program was designed to be a media specific, rather than a pollutant specific, approach. Individual countries took responsibility for developing individual sections: atmosphere (Canada), terrestrial (Sweden), freshwater (Finland), marine (Norway), human health (Denmark), remote sensing and modeling (the United States) and emissions and discharges (the secretariat). In the preparatory process, the AEPS participants recognized that two of the most significant threats to the present Arctic environment may come from climate change and stratospheric ozone depletion. However, programs to study these problems have been developed in other international fora, therefore they were not included in the Rovaniemi Process in 1991. These global problems were included on the list of AMAP activities at the 1993 Nuuk Meeting (The Arctic Environment, 1993).

Organizing the collection of the information was complicated. The 1993 AMAP update noted that 'a combination of lack of data and restrictions on access to data' imposed by various commercial and military authorities, has made the work of AMAP difficult. Thus, the AMAP has been unable to provide the AEPS meetings with comprehensive emissions and discharge data. According to the update, 'this situation makes it impossible to compare the contributions and significance of Arctic versus non-Arctic sources and thus recommend comprehensive, optimal response strategies for the entire Arctic at this time' (AMAP, 1993a, p. 8). The progress made within the AMAP has been slow because of several hindrances: 1) the financial base of the program has not been strong, 2) the differences in monitoring data and the practices of collecting and handling it vary according to regions and scientific traditions, and 3) some information has not been made available although it had been promised. Interestingly, cultural differences in monitoring and research practices between different regions and institutions, rather than the financial basis of cooperation, seemed the main reason for the slow process of collecting the monitoring data (Nenonen, 1996).

The problems were known early. For example the International Arctic Science Committee (IASC), with the International Council for the Exploration of the Seas (ICES), was asked by the AMAP to perform an audit on the monitoring program of the first AMAP period (1992-1996). Independent experts performed the audit. The auditors reported many problems in the work of the AMAP. The main conclusions were, according to the IASC audit of 1993, that relying on the existing programs will not suffice to achieve AMAP objectives. The relationships among the five sub-programs were not sufficiently coordinated or harmonized. The report criticized the plan because it was difficult to derive a holistic picture of Arctic contaminant transport, dissemination and deposition. The decision by the AMAP task force to monitor along media-specific boundaries instead of pollution-specific lines was considered to give reason for concern. The relationship between national programs and the overall implementation plan of AMAP was considered weak. No member country has attempted to describe the assessment of the AMAP. The auditors evaluated that the time frame of AMAP report was unrealistic (AMAP, 1993b).

This puts pressure on the assessment work of AMAP; assessment means interpreting existing information for the SAAO's and ministers and making suggestions for further measures by the states nationally, regionally and internationally (see also AMAP, 1995). The assessment task of the AMAP, resulting in the 1997 report, has been viewed 'a considerable burden' on the AMAP in negotiating meaningful policy recommendations within an organization by personnel with a scientific rather than a policy portfolio (Ringold, 1994, p.103). The assessment report on the status of the Arctic environment was delayed. The assessment was not really based on the latest information originating from the AMAP activities; a major part of the AMAP assessment was based on information and results published in scientific literature and was available in scientific reports. The AMAP stressed that an assessment utilizes accumulated data and information and does not necessarily require any new data to be collected during the assessment process (Reiersen, 1996).

The 1997 report by AMAP suggests existing legal regulations are followed and reviewed to ensure full accounting for the extreme Arctic conditions. These measures, however, are not enough since levels of many contaminants in the Arctic are, according to the report, likely to remain at or close to existing levels for decades because of their resistance to degradation, the slow rate of degeneration and the recycling of existing accumulations. Thus, ameliorative action to reduce exposure to humans and to protect wildlife is 'essential' (AMAP, 1997, p.vii).

Despite these problems, the AMAP report constitutes the Arctic as a region for a moderate environmental concern. Compared to most other areas of the world 'the Arctic remains a clean environment'. Some pollutants, and the combinations of different factors, however, 'cause concern in certain ecosystems and for some human populations' (AMAP, 1997, p.vii). The constitutive nature of knowledge emerges when knowledge is understood as more than scientific knowledge on emission sources, levels and their ecological, economic and social implications. For environmental cooperation, knowledge of pollutants and anthropogenic emissions is an important part of the management of the problems but there is more to it than this instrumental knowledge. Foucault (1972, p.185) combines 'savoir' with knowledge instead of 'connaissance' as knowledge; the difference between connaissance and savoir is the difference between instrumental and constitutive knowledge. The example is easy to find; scientific knowledge (connaissance) of mental diseases plays a role in the knowledge (savoir) of madness (a question of decision-making authority). By concentrating on this constitutive aspect of knowledge the subject and the object, and the knowledge about them, is mutually constitutive.

It seems that the knowledge acquired by the AMAP legitimizes the existing situation and the lack of special measures to protect the Arctic environment. One of the close observers to the process, the AMAP was not established because the concern for nature. The proposal for the AMAP was made knowing the rich natural resources that the area has. Then, however, the 'general opinion' was 'environmentally active'. Establishing AMAP was a sign of 'finding' the Arctic but there were parties in the negotiations, in particular Norway, that did not like the purely environmental protection initiative. The rationale for establishing AMAP, in which Norway took a leading role, was that 'if there would not be problems ahead in the future, why would you need a monitoring program?'. The AMAP was established because the region and its resources would be used and therefore the states needed a mechanism of surveillance to make sure that any crucial limits were not trespassed (Nenonen, 1996). According to the AMAP, the region is now monitored so that should some adversary developments suddenly take place information about it could be collected throughout the circumpolar region (AMAP, 1997, p.2).

Other Reports

Besides the AMAP report, other reports by different institutions have been published in the last two years. In a report by the European Environment Agency the Arctic environmental threats are 'regional'. This report, partly based on information produced within the AEPS, defines eleven main current threats to the European Arctic environment. They are:

> Habitat fragmentation, degradation, or destruction; over harvesting of biological resources; the potential or radioactive contamination; persistent organic pollutants; oil pollution; tourism in vulnerable areas; introduction of alien species and diseases; cumulative impacts; long-range pollution transport; climate change; ozone depletion and UV radiation (Hansen et al., 1996, p.113).

The report stresses that one of the 'most valuable and unique features' of the European Arctic is its large area of wilderness and natural wildlife habitats. Agriculture, forestry, and other natural resource exploitation, road, construction, urbanization and other infrastructure development are all human activities that cause habitat destruction and fragmentation. Such activities continuously and increasingly affect wildlife habitats in the European Arctic. Several mammal and bird species have previously been over harvested in the European Arctic and both fish stocks and forest areas are being harvested today (Hansen et al., 1996, p.113).

A third report on the state of the Arctic environment was produced by the Nordic Council of Ministers in 1996. It deals with Lapland in Finland, Norrbotten and Västerbotten in Sweden, northern Norway, Iceland, Greenland and Svalbard. The report is interesting since it aims to rank environmental concerns. The writers have attempted to analyze the environmental impact of both the use of natural resources in the region and the threat caused by pollution (Bernes, 1996).

The report deals with biodiversity and concludes that, rather than pollution, the resources and recreational values of nature in the Nordic Arctic region dominate concern. The impact of natural resources' exploitation is a threat to the biodiversity of the region. The report states that the greatest threats to biodiversity are: hunting, climate change, overgrazing, forest fires, hydroelectric power, the introduction of new species, erosion and earlier deforestation. The threats to the use of natural resources are: the use of minerals and fossil reserves, overfishing, deforestation, hunting and overgrazing. The greatest threats to the recreational value of the environment are, however, lower than in the previous cases: climate change, overgrazing, erosion, water energy, noise and tree felling in the fjells (Bernes, 1996, p.233). The report discusses the issue of pollution to some length. It says that local pollution has had severe impacts on these values and some pollutants are spread over the Arctic. None of these pollution effects have, however, been both large scale and severe on the Arctic ecosystems; at least at the moment no such observations can be made on this point. The health risks are, however, connected to the pollution problem in the region but no indications have been found to indicate that these pollution effects would be lethal (Bernes, 1996, 233).

The hunting families of Greenland are singled out as being threatened by pollution because of their diet. The report considers the dominance of the pollution problem in the public sphere and shows that such problems are easy to blame for the degradation of the environment. The problem is outside the hands of local people and has to be solved somewhere else. The issues related to the use of natural resources, and threats to them, direct discussion to ways of living and consumption and the need to change them locally. Such discussions are not easy (Bernes, 1996, p.233).

In addition, results of two national research projects have become available. The basis for the Finnish Forest Decline Project was the awareness of extremely high emissions of sulphur dioxide and heavy metal and severe environmental degradation in the Kola peninsula. A connection between these emissions and the needle-loss outbreak that occurred in Salla, eastern Lapland, was suspected. Between 1990 and 1994 fifty researchers from Norway, Finland and Russia studied the problem. The studies were concentrated on a series of joint sampling and monitoring plots located along lines running through Finnish Lapland to the northwest, west and southwest, from both Nikel and Monchegorsk in the Kola peninsula (Tikkanen, 1995; see also Väliverronen, 1996).

The report concludes that there were considerable emissions from the Kola peninsula arriving in eastern Lapland on easterly and north easterly winds. The duration of these high pollution episodes, however, is short and the annual mean sulphur dioxide concentration in Lapland is relatively low. The project was not able to demonstrate any direct connection between the cases of tree damage reported at the end of 1980's in Finnish Lapland and emissions from the Kola Peninsula. However, sensitive 'bioindicators,' such as lichens and the sulphur and acid rain damage symptoms in pine needles, indicated that the emissions are having an effect in Finnish Lapland (Tikkanen, 1995; p.211-216).

A Canadian review team considered nine organochlorine compounds and three metals in their Northern Contaminants Report. This stated that 'contamination of traditional foods does not pose any immediate short-term health threat'. In the Canadian north, people who eat a lot of marine mammals tend to have higher contaminant levels than at 'zero risk' level, but more research needs to be undertaken to understand the long-term health issues. The report concludes that 'the nutritional advantages of traditional foods still outweigh any potential risk which may be associate with them'. The report stresses that hunting and consumption of wildlife, fish and plants are essential to the health, culture, spirituality and economy of the people of the north. If people stop eating traditional foods, which may have many overriding benefits, adverse effects may occur (Northern Contaminants Program, 1997, p.7-8). Both reports conclude that the current levels of exposure to contaminants in the Arctic are clearly of concern but it is still not clear what public health measures should be taken (see also AMAP, 1997, p.186).

The relation between these reports is worth examining more closely. The point of having these reports on the state of the Arctic environment is that they 'balance the information about the Arctic' (Liljelund, 1996). They do so by increasing information about the knowledge of threats other than those related to the long-range transportation of pollutants to the Arctic. Both the report of the European Environment Agency and the Nordic report emphasize the threats to the biodiversity of these northern regions, compared with the AMAP emphasis on global, regional and local pollution issues. The Finnish and Canadian reports downplay the threat of pollution in the Arctic. In conclusion, there is no consensus on the state of the Arctic environment and threats to it outside the AMAP.

Arctic Science and Indigenous Peoples

Scientists and the Arctic Cooperation

Sweden stressed the need to improved scientific cooperation during the preparatory process of the AEPS:

> Science will help us identify the threats to the environment. Science will also aid in designing appropriate counter measures. We have to make sure that the scientists have the opportunity to carry out their important activities (Statement of Swedish Delegation, 1989, p.5).

The role of science in Arctic environmental cooperation was emphasized by Fred Roots, representative of IASC at the Rovaniemi ministerial meeting in 1991:

> Wise actions and policies to protect or maintain life quality and productivity of the environment... can only be achieved if there is adequate knowledge on the natural characteristics and processes of the Arctic regions (Roots, 1991, p.1).

To Roots, science is 'the organized and systematic pursuit, accumulation, and testing of observations, data, experience and insight, and turning that information into knowledge that can be disseminated for all to use' (Roots, 1991, p.2). The IASC offered its services to the Arctic states at the meeting in Rovaniemi. The IASC is a nongovernmental multisubject scientific organization whose purpose is to facilitate international cooperation and the exchange of information for all kinds of scientific activity in the Arctic regions. The committee does not conduct research; it has the responsibility of developing policies and guidelines for cooperative scientific research concerned with the Arctic (IASC Founding Articles, 1990, p.1; see Roots, 1992, 1993 for a history of the IASC).

Over the years the IASC has developed 'an informal relationship' with the AMAP (IASC Council Meeting, 1992, p.3). As the only accredited scientific observer to the AEPS, there were discussions of the role of the IASC in the framework of the Arctic Council. In principle the scientific community views the establishment of the Arctic Council in positive light and looks forward to the development of the relationship between them: 'The Arctic Council promises to provide the political framework within which to establish science politics and priorities, and thus provide a rational basis for science resource allocation in the Arctic nations'. At the IASC Conference in 1995 the match between the Arctic Council and IASC activities was described as 'excellent' by one of the key note speakers. In 1996 IASC Council meeting it was decided that IASC would seek an advisory role in the Arctic Council (IASC Meeting, 1996, p.38).

However, the questions that need answers are: Who speaks for Arctic science? How do governments obtain and use international science policy advice? How can science systematically identify and solve circumpolar problems? What is the role of government research agencies in Arctic science? And, To whom are Arctic scientists responsible? (Lock, 1995, p.12).

The role of the IASC in relation to the Arctic Council is defined as either a facilitator or an advisor, according to Lock. However, 'the responsibility for science policy lies, ultimately with the political community' (Lock, 1995, p.15). The Arctic Council could articulate or imply Arctic science policies rooted in national policies. These discussions are 'political, not scientific'. In Lock's view, science is incorporated only at the early stage, during international discussions. Under these conditions, advice from an international body such as the IASC would be 'superfluous at best; at worst, it may prove wasteful and vexatious, especially where the science politics or priorities of the Arctic nations are not in accord' (Lock, 1995, p.16-17).

Alternatively, the Arctic Council could choose to seek scientific advice from an independent circumpolar body set up especially for that purpose. Such a body 'could do much to smooth the development of circumpolar policies'. This would not optimize the chances of reconciling the policies of individual Arctic nations but would, or could, facilitate appropriate contributions from non-Arctic nations. The functioning of such an advisory body would, however, require it to be independent of any organization that has interest in its advice. Any scientific bodies concerned with the funding and execution of Arctic science would have to remain separate from this advisory body. 'In short, those that do the science, or have it done, should not be those who recommend it,' concludes Lock (1995, p.17).

Lock suggested the establishment of a science and technology advisor to the Arctic Council named Arctic Science Advisory Board. Oran Young representing the IASC voiced the concerns of the scientific community about such a body in the AEPS meeting in Alta, Norway:

> Our experience with scientific and technical committees built into international agreements is not encouraging. The danger of politization is ever present when the connection between science and policy becomes too close (Young, 1997, p.1).

According to this view, the policy community must be free to seek relevant scientific information and advice from a variety of sources including, but not limited to, the IASC. On the other hand, the IASC must be free to develop a science program that addresses 'the cutting edge questions of concern' to scientists working in this field. In this sense Young thinks that the AEPS got it right in 1991 in saying 'the eight Arctic countries will consult, as deemed appropriate, with the International Arctic Science Committee and other bodies on any matter that falls within the scope of this Strategy'. Young (1997, p.4) hopes for 'an easygoing relationship between the two communities in which each is able to assist the other and neither compromises the integrity'. The U.S. SAO (Senior Arctic Official) saw the role of IASC and the science community as important in the Arctic cooperation: science may be an 'initiator', but often also 'a marker', or milestone by which policies can be evaluated. In the Arctic context, the range of science issues 'go far beyond those traditionally imagined as being arctic science' (Senseney, 1997, p.4).

However, in a recent evaluation of the AEPS, it was noted that 'scientific information is not always delivered to the SAO's in an easily grasped from, thus making the policy formulation process hard to manage'. Also, the capacity to handle all the information made available seems to have reached 'its maximum limit'. All of the available information is not used nor did any party or a person seems to have a holistic view of the whole information base (Nilson, 1997, p.5).

As a solution to these problems, the report suggests the establishment of an advisory board or a screening function. An advisory body could be either scientific or technical in its nature. A purely scientific advisory board probably would take holistic approach to matters compared with a more technically oriented advisory board staffed by government agency experts. A screening body could be staffed with knowledgeable people who can advise the SAO's in their decision-making. It could help in communication between the SAO's and the working groups. Either a screening body or a technical advisory group probably could function as a translator and a filter between the SAO's and the working groups (Nilson, 1997, p.29).

The role of scientists in such a situation, however, creates different views. How far can scientists go in interpreting research results? Scientists have power in that they are in control of problem framing. Epistemic communities are networks of knowledge-based experts which articulate 'the cause and effect relationships of complex problems, helping states identify their interests, framing the issues for collective debate, proposing scientific policies and identifying salient points for negotiation' (Haas, 1992a, p.2; see also, Haas 1992b).

The mainstream regime approach assumes that knowledge arises as an issue in the absence of power relations. The epistemic community theorists, for example, focus on uncertainty because they assume that in such a situation the power aspect may be absent and institutions work poorly. It is assumed, however, that epistemic communities do have the power to create reality but 'not as they wish,' since political factors and related considerations affect the construction of the reality just as much as the contribution of scientific experts (Adler and Haas, 1992, p.381).

The members of epistemic communities do not have an equal access in political decision-making. Steinar Andresen et al. (1994, p.117) emphasize the role of 'those scientists who are most involved in the dialogue with decision-makers and serve as coordinators of research and as mediating agents between those who do the actual research and the decision-makers'.

Karen Litfin has created a concept 'knowledge brokers' to capture the elements missed by the epistemic communities' theorists. By knowledge brokers she refers to those intermediaries between the original researches, or the producers of knowledge, and the policy makers who consume that knowledge but lack the time and training necessary to absorb the original information. The ability of knowledge brokers, who operate at low or middle levels of governments of international organizations, to frame and interpret scientific knowledge is a substantial source of political power, according to Litfin (1994, p.4).

The approach by Litfin focuses more on individuals and their role in the interface between science and politics than groups such as epistemic communities. According to this view, 'interpreting and framing knowledge become crucial political problems as information is mustered to achieve policy objectives' (Litfin, 1994, p.8). Individuals and groups function as creators of risk and carriers of knowledge. This means that knowledge is embodied, that is concrete individuals and groups of individuals who serve as definers of reality. To understand the role of the knowledge brokers is to situate them in a discourse, which also means that subjects are at least partially constituted by the discursive practices and contexts in which they are embedded. The knowledge-broker who interprets and frames knowledge for policy-makers has power in respect to 'how knowledge is framed, by whom and on behalf of what interests' (Litfin, 1994, p.198).

Controversial Relationship

The relationships between science, knowledge and indigenous peoples in the AEPS were, from the indigenous peoples' point of view, one of the main concerns. The relationship between indigenous peoples and science is 'complex and often controversial'. It is controversial because indigenous peoples are concerned about who actually controls research. They also question whether scientific studies are beneficial to their own society, economy and environment. Science has the power to validate the opinions of outside interests in circumpolar issues (Brooke, 1993, p.17).

The problem is based on misunderstandings between indigenous peoples and the scientific, administrative and political interests of the member countries. These misunderstandings are based on: 1) resistance by member countries to share the responsibility for planning and decision-making, 2) there is resistance by scientists and administrators concerning the role, content and utility of indigenous knowledge, and 3) there is resistance by many indigenous peoples to the value and accuracy of scientific information and scepticism about the motives that underlie international processes such as the AEPS. Indigenous peoples stress that research activities and the findings and opinions of scientists are import aspects of decision-making in the southern centers (Brooke, 1993, p.12; see Brelsford, 1995; Brelsford, 1996b; Cochran, 1996). Indigenous peoples consider that the Arctic scientific community, for example the IASC, has not been very receptive to the idea of including the knowledge of the indigenous peoples. The IASC saw itself as 'a science organization': it did not want to become 'a tool for social or policy action'. Relating this to the inclusion of human and social sciences has been problematic in the IASC. In 1991, it was recognized that the IASC itself should have an important future role in binding links between natural and social sciences but such a role would have to be developed carefully and with 'considerable sophistication and sensibility' (IASC Council Meeting, 1991, p.11).

There was a proposal that a standing working group could be established to represent the interests of social, human and medical sciences. It was not, however, adopted. The problem, from the IASC standpoint, seemed to be one of no well developed criteria for what is excellent, or leading, in the social or human sciences (IASC Council Meeting, 1991, p.11). The IASC Council decided an advisory group that would include representatives from IASC Council, International Arctic Social Sciences Association (IASSA, established just before IASC) and indigenous peoples of the north should be established. The advisory group would include three representatives of IASSA, three representatives from the major indigenous organizations and two members of IASC member organizations (IASC Council Meeting, 1993, p.9). The IASSA response to this proposal was negative and the organization proposed becoming an advisory group to the IASC on social sciences, as had been done with the International Union on Circumpolar Health (IUCH) in 1993. The IASSA was invited to become an advisory group to IASC in 1994 and the contact persons were nominated (IASC Council Meeting, 1994, p.13).

However, this solution left open the issue of participation by indigenous peoples. Franklyn Griffiths, during his speech at the IASC Council Meeting in 1995, questioned some approaches adopted by IASC; the way, for example, the IASC had been involved in 'mega-science' and 'under performing' the residents of the area. He raised the question of whether indigenous peoples or users of Arctic science should be directly represented on the regional board of the IASC. As to making Arctic science relevant, he saw no other alternative than making IASC programmes as socially responsive as possible (IASC Council Meeting, 1995, p.15).

The problem of participation by indigenous peoples in IASC activities was seen also at the IASC Arctic Research Planning Conference in Hanover, and in the United States, in December 1995. Addressing the IASC conference, Canadian Arctic Ambassador Mary Simon stressed that the Arctic national governments have agreed upon the importance of indigenous peoples' knowledge and their contribution. This, she said, was also expected to happen among the larger scientific community (Simon, 1995). Tove Petersen from the Indigenous Peoples Secretariat (IPS) pointed out at the Tromsø Conference on Environmental Pollution in the Arctic in June, where the AMAP assessment report was published: 'a new form of colonialism is created if there is no communication between scientists and indigenous peoples'. Unless there is communication between different groups, the production of knowledge in the Arctic is an 'affirmation of existing power structure' (Petersen, 1997).

Communication in such a situation needs to be analyzed. Those who emphasize the idea of a communicative rationality focus on an ideal speech situation where there are no hindrances to communication. Although actual speech situations hardly ever resemble the ideal situation, these basic assumptions make it possible to define the truth as 'a rational consensus'. The basis of a situation is characterized as pure intersubjectivity by the absence of any barrier which would hinder communication (McNay, 1992, p.183). International environmental politics have been studied with the help of the idea of communicative rationality (Dryzek, 1990; Hjorth, 1992).

Ernst B. Haas (1990, p.11; 20-21) emphasizes that consensual knowledge is always socially constructed and therefore 'inseparable from the vagaries of human communication'. Haas claims that 'change in human aspirations and human institutions over long periods is caused mostly by the way knowledge about nature and about society is married to political interests and objectives'. Knowledge is 'a social product' which denotes that it is easily politicized by becoming a field of struggle in itself. Neither the processes of making knowledge more extensive nor the reflection on knowledge itself are passive. They are 'intensively political' (Kratochwil and Ruggie, 1986, p.773).

The idea of communicative rationality is utopian, according to Foucault, because of the relations of power in society: 'the problem is not trying to dissolve them (relations of power) in the utopia of a perfectly transparent communication, but to give one's self the rules of law ... which would allow these games of power to be played with a minimum of domination' (In, McNay, 1992, p.184).

The Foucaultian view on knowledge considers it as discourse. Knowledge, is a system for the formation of statements. Knowledge is 'that of which one can speak in a discursive practice and which is specified by that fact' (Foucault, 1972, p.182). According to Foucault, 'we should abandon a whole tradition that allows us to imagine that knowledge can only exist where the power relations are suspended and that knowledge can develop only outside its injunctions, its demands and its interests' (In, Simons 1995, p.44). Foucault questions the idea that knowledge is power or vice versa; the relationships between power and knowledge are complex and not to be taken for granted. The relationships are overlapping, reciprocally supporting each other, using and contradicting each other (Foucault, 1982a, p.217-218). It follows that discourse is '...by nature, the object of a struggle, a political struggle' (Foucault, 1972, p.120).

Foucault points out that studying the interactions between knowledge and power cannot be studied at the discourse level by analyzing concepts, subjects, objects, enunciative modalities and strategies. The institutions of power/knowledge have to be used as the basis for analysis. Foucault takes the language-politics connection to a higher level of abstraction from linguistically reflected power exchanges between persons and groups, to an analysis of the structures within which they are deployed. He argues that the deep-seated discursive formations which determine the production of knowledge in a given period are intimately bound up with nondiscursive factors defined as 'an institutional field, a set of events, practices and political decisions' (Foucault, 1972, p.157).

Indigenous Knowledge in the AEPS

In 1989, the Danish had already suggested the use of indigenous knowledge as a part of Arctic environmental cooperation in the AEPS: 'their traditional knowledge is taken into account on par with otherwise accepted knowledge' (Statement by the Danish Delegation, 1989, p.3). In the declaration signed in 1991, the states emphasized recognizing 'the special relationship of the indigenous peoples and local populations to the Arctic and their unique contribution to the protection of the Arctic environment' (Declaration on the Protection of the Arctic Environment, 1991). It was recognized that 'this strategy must incorporate the knowledge and cultures of indigenous peoples' (AEPS 1991, p.2). In practical terms, the 'consideration of the health, social, economic and cultural needs of indigenous peoples shall be incorporated into management, planning and development activities' (AEPS 1991, p.6).

Two decisions were made in Nuuk 1993 to enhance the participation and contribution of indigenous peoples. First, the government of Iceland offered to host a seminar on indigenous knowledge. A seminar was held in 1994 to clarify how indigenous knowledge was applicable to the AEPS and its programs. It aimed to identify strategies and pragmatic proposals for the integration of indigenous knowledge into the AEPS and its programs and to identify the contribution that indigenous knowledge could make to sustainable development. Second, the Indigenous Peoples Secretariat (IPS) was established with the financial help of Denmark to assist and coordinate cooperation between indigenous peoples' organizations (The Arctic Environment, 1993).

The idea in the AEPS is that 'knowledge is in people' (Huntington, 1996). From the indigenous point of view, indigenous knowledge is 'a body of knowledge on its own right and a means of communication and decision-making that reflects who indigenous peoples are and the world view that they hold' (Brooke, 1993, p.75). Indigenous knowledge is, according to the AEPS seminar report, considered a 'legitimate, appropriate and valuable basis for co-management and self-determination'. It is dynamic and evolves in response to changing situations and conditions. Moreover, there is 'an inseparable relationship' between the utilization of living resources, conservation and the continuing vitality and utility of indigenous knowledge. The AEPS seminar saw the role of indigenous knowledge as 'relevant'. It could help identify important research areas, expand understanding of the natural world, and bring useful insights into natural processes - including the role of humans in the environment. At the opening of the AEPS seminar on indigenous knowledge many participants stressed that the issue was not 'integration but partnership' (Hansen, 1994, p.16-17).

The knowledge of local and indigenous peoples is not usually evaluated as 'knowledge'; in some instances, it has been part of the fieldwork of a western educated scientist. It is rarely authorized as knowledge, there is no formal external science for acquiring that knowledge and the status that come with it. It is not transnational, nor is it a network. It is local and piecemeal information about the human-nature relationship in certain cultures and by certain persons. This knowledge includes empirical or practical knowledge. It also includes paradigmatic knowledge, for example about the way to interpret practical knowledge and construct mostly coherent cosmologies. Finally, it is institutional, since it refers to participation to decision-making (Kalland, 1994, p.157-160).

The knowledge of indigenous peoples is often defined as 'traditional knowledge'. Traditional knowledge is not based on the formal methods of scientific knowledge. It is passed through socialization from one generation to another and by the involvement of children in everyday life activities. It includes the names of places and creatures in nature, and they do not necessarily follow western logic nor western ways of mapping. It is not formal, in that there are no courses nor diplomas to be taken in traditional knowledge. Traditional knowledge is considered endangered because the people who have that knowledge and their lifestyles are changing (Freeman and Carbyn, 1988;

Freeman, 1992; Inglis, 1993; Borgos, 1993).

It is sometimes stressed that this knowledge is connected to the environment. Claiming that it is 'traditional ecological knowledge' (TEK) refers to the paradigmatic nature of information, the way of interpreting practical knowledge. It constructs its own cosmologies. This encompasses spiritual relationships, relationships with the natural environment and the use of natural resources, and the relationships between people. It is reflected in language, social organization, values, and the institutions and laws of indigenous peoples (In, Bell, 1994, p.190). This knowledge is seen as an alternative to western scientific knowledge because of the differences in their world view. Henry Huntington defines traditional ecological knowledge as a 'coherent world view with internal and external relationships that may not have direct parallels in western science'. TEK typically embodies 'a holistic view of ecology rather than an atomized one and some understanding of the local taxonomy is necessary to understand TEK fully' (Huntington, 1994, p.89).

Indigenous peoples themselves have argued against the term 'traditional' since it does not consider that this form of knowledge is evolving and dynamic which adapts to changing conditions and situations. Some would leave the environment out, and concentrate on 'traditional knowledge systems' as suggested by Elina Helander. To her this knowledge is, 'patterned ways in which peoples from a non-literate tradition learn about their reality and communicate such information amongst themselves and from generation to generation' (Helander, 1993, p.71; see also Eidheim, 1995; Helander, 1996).

Sometimes, particularly in relation to the environment, this knowledge is identified as 'local'. It is connected to a certain population and place: Indigenous knowledge is considered dependent on the existence of indigenous peoples' and their way of life. Indigenous knowledge is the foundation of a culture in its own right. The expertise cannot be separated from the practices and experiences of particular local, cultural groups (Hansen, 1994, p.17). Some critics claim that this knowledge is not valuable because it is held by a certain group of people such as indigenous peoples. Its value is in what it has to say about the ecosystem and place of people within it. The practical value of this knowledge is most important: 'local knowledge is too important to be tamed into myths for the environmentalist consumption' (Kalland, 1994, p.152, 157).

In this work, 'indigenous knowledge' is understood, following the ICC report on indigenous knowledge, to comprise:

> ... information and concepts about the environment and ecology that are known but usually not formally recorded by individuals who belong to a particular cultural group that has occupied an identifiable territory over a long period of time (Brooke, 1993, p.36).

It includes 'facts, concepts and theories about the characteristics which describe the objects, events, behaviors and interconnections that comprise both the animate and inanimate environments of indigenous peoples'. This knowledge is local since it has been developed 'through the person's observations of, experience with, and explanations about the physical environment and living resources that characterized the territory in which they live'. This knowledge is traditional because it is 'commonly shared between individuals' and it is transferred from one generation to the next through the oral tradition. The content and extent of knowledge 'vary from individual to individual and there can be a specialization in expertise'. It is also paradigmatic: All of this information is organized around concepts and perceptions that are 'constantly being shaped and reshaped by the intellectual culture of indigenous peoples and its content and meaning is best expressed within the context of indigenous language systems'. It also has the capacity to provide 'explanations about causality and give validity to the world of natural phenomena' in a way that is consistent with systems of belief and which characterized the world view of each indigenous society (Brooke, 1993, p.36-37).

Indigenous Peoples and Knowledge

Knowledge in Regimes

The question of indigenous knowledge is about power; demanding the use of indigenous knowledge is a demand that 'the power base must be shared' (Brooke, 1993, p.18). Therefore, this knowledge is also institutional, since it refers to local participation in decision-making. The growing awareness of different types of knowledge can be seen both as a response to the need of acknowledging the role of local informants in producing knowledge about the northern regions and the demands by the local communities for researchers to acknowledge their contribution to western science. The aim is to use different kinds of knowledge 'to present different sides of problems and advance creative thinking'. The aim is not to find 'an average for different views' (Hilden, 1996).

The institutional dimension means that members of indigenous peoples organizations are experts whose knowledge and expertise can and will be used in the AEPS. By claiming the knowledge is 'indigenous' the basis of using this knowledge becomes political. This means that the important point in this knowledge is not necessary the information content but the participation of indigenous peoples as experts in cooperation. Stressing 'indigenous' knowledge as opposed to traditional, ecological or local, makes the issue into an institutional problem, a problem of participation in decision-making. It refers to the direct participation of indigenous peoples' groups in decision-making. According to the indigenous peoples, the use of their knowledge is an issue of self-determination (Hansen, 1994, p.16).

These processes of inclusion and exclusion raise the questions of who has the right to speak, from which institutional places speeches can be made and the kind of positions the objects of the discourse can have. Initially, this means that one has to describe the institutional sites from which the speakers make their discourses and the legitimate source and point of its application (Foucault, 1972, p.51-52). The question Foucault asks is of the relationship between knowledge and political practice. The various status, the various sites, the various positions that he can occupy or be given when making a discourse is interesting. The definitions of positions include institutionalized justification:

> The method of justification and refutation confers on these speech acts their claim to be knowledge (savoir), and makes of them objects to be studied, repeated and passed on to others (Dreyfus and Rabinow, 1982, p.48).

According to Foucault, a rival knowledge to that of scientific knowledge may exist; the hegemonic discourse and its disciplines may be unable to absorb or obliterate these fully and may even have to reach an understanding with some of them. Alternative knowledge combined with local power may defy the dominant discourse and its apparatus (Keeley, 1990, p.94). This alternative knowledge is a range of formulations that either have never gained formal recognition as regimes of truth or lost that status; they are 'subjugated knowledge'. Subjugated knowledge is historical content that has 'been buried and disguised in a functionalist coherence of formal systemisation'. Subjugated knowledge is thus 'blocks of historical knowledge'. On the other hand, subjugated knowledge is 'a whole set of knowledge that have been disqualified as inadequate to their task or insufficiently elaborated'. Foucault calls these blocs popular knowledge, though it is far from being general common sense knowledge, but is a particular, local, regional knowledge (Foucault, 1980a, p.81-82).

Indigenous Expertise

Many of the projects that aim to include indigenous knowledge are still in progress and writing any final evaluation of their significance and results is therefore difficult. It is clear that the role of indigenous knowledge has been recognized although this recognition varies between different working groups.

The goal of the fourth program under CAFF, the integration of indigenous peoples and their knowledge, is to 'fully integrate indigenous peoples and their knowledge into the functions, processes, and implementation of CAFF'. This program includes four subprograms: the Indigenous Knowledge Mapping Project on the Beluga Whale, Ethical Principles for Arctic Research, the Indigenous Knowledge Data Directory and the Review of Co-management Systems (CAFF, 1995-1996, p.9). The open policy of CAFF to indigenous peoples organizations and their participation in different meetings caused some confusion in the Moscow CAFF meeting in the fall of 1995 (Hild, 1996).

There is a project within the CAFF that reflects the concerns of the indigenous peoples towards scientific research in the northern regions. The United States has completed a compilation of ethical principles from various sources for Arctic research. The SAAO's forwarded the review to IASC which coordinates with the IASSA in their preparation of pan-Arctic ethical principles. The projects, under the fourth item of the CAFF Work plan, involve Inuits from Alaska and Canada and the Alaskan-based Arctic Network. Under Canadian leadership the ICC is preparing a data directory for locating sources of indigenous knowledge for the Arctic (CAFF, 1995).

Another project focuses on the traditional ecological knowledge of the Beluga Whales and aims at demonstrating an appropriate methodology for documenting indigenous ecological knowledge. It is to recommend appropriate steps for integrating this knowledge into the work and policies of the AEPS. Beluga whales were selected as the target species for the pilot project because they migrate across national boundaries and are of scientific interest (see Huntington, 1994).

The 'Ice-Edge Ecosystem Project' of the Arctic Network aims to identify and map areas in the ice edge environment which are important for high biological diversity, biological activity, large seasonal concentrations of species, and for cultural and subsistence purposes. These maps are to be based on traditional knowledge and existing scientific data. The project also aims to identify and set priorities for the ecological health of the ice edge environment, the future of traditional harvesting and subsistence life and to encourage a cooperative effort among native and environmental groups to protect habitat and species' diversity for continued subsistence purposes (Gibson, 1996).

The CAFF program suggested that co-management as a model for the effective participation of indigenous peoples in resource management needed to be explored. Both Canada and the United States are jointly leading a review of co-management systems to analyze the strengths and weaknesses of the mechanisms for involving indigenous peoples in Arctic resource management. The AEPS seminar report point outs that if the co-management of renewable resources was to be effective, formal agreements between indigenous peoples and governments were required. To address their need for hunting rights, indigenous peoples require territorial rights and participation in the management of land and living resources (Hansen, 1994, p.19; see also CAFF, 1995, p.3).

Indigenous peoples' organizations have been represented at EPPR official meetings (Pahkala, 1996). The ICC has expressed its satisfaction over the establishment of cooperation between itself and the EPPR but has pointed out existing communication problems.The need of communication was stressed; The ICC indicated that indigenous peoples should be involved in identifying areas that are biologically and culturally sensitive and areas for hunting and traveling. Indigenous peoples should receive information about emergencies in terms and languages understandable to them (see ICC, 1995a) (EPPR, 1996, p.15).

With PAME the role of indigenous peoples and their knowledge has been very modest. Indigenous peoples' organizations have been present at the meetings of PAME. The 'ICC and Sámi Council have been more visible than the Russian association'. However, 'the nature of cooperation within PAME does not require the use of indigenous peoples' knowledge'. Though, it is 'important is that they are present in the meetings' (Lassig, 1996).

The evaluation made by the IPS in 1996, during the earliest phase of cooperation in 1991-1993, stated: 'many working group meetings were unattended or the indigenous peoples' representatives were ill-prepared'. The organizations often did not receive notices of meetings and not much information on AEPS activities was distributed to the indigenous peoples organizations. The funding to support the participation of the indigenous peoples was 'minimal:' the government of Canada, and to a lesser degree Denmark, provided the majority of funds during this period and this explained in part why the Inuit were more often represented than other indigenous peoples. The report stresses that most of the funds allocated during the 1991-1993 period were primarily for meeting participation. Very little financing was made available for meeting preparation or for communicating the work of the AEP's to the organizations' own people, whom they represented (Reimer, 1996, p.3).

Since 1993, the ministers have given the three indigenous peoples organizations a formal standing invitation to all future meetings of the AEPS, making it 'a milestone'. Between 1993 and 1996 there was a noticeable increase in the participation of the three indigenous peoples' organizations. The SAAO meetings were generally 'well-attended' by representatives of the three organizations. Additionally, AEPS working group meetings were also better attended than during the earlier period. However, excepting ICC, the working group meetings were often attended by volunteers or others who were not well prepared. The degree to which meaningful input was made directly related to the funding available for preparing the issues beforehand and to discussing the issues 'back at home' with the organizations' own people. Chester Reimer, the executive director of the IPS, described the form of participation in the later period as 'project participation'. Within each of the working groups several tasks and projects were undertaken. Sometimes, projects were led by an indigenous peoples' organization, such as with the sealing industry project by the ICC. In other cases, parts of reports were contracted out to indigenous peoples' organizations, such as sections of the AMAP Assessment Report. In still other cases, indigenous groups were partners in co-leading a specific project (Reimer, 1996, p.3).

There is a rather large consensus among the participants that indigenous involvement in the AEPS has made the process 'a different product,' compared with their lack of participation (Whitby, 1996). It provides 'real life examples'. The 'indigenous peoples' role is both important and positive in the Arctic,' according to the head of the U.S. delegation, Robert Senseney (1996). For the environmental coordinator of the ICC, the current situation is best described as, 'a means to an end... to get more influence and control'. The benefits of indigenous knowledge are many: 1) 'to make a real contribution,' 2) 'a nice, interesting new thing to attract funding,' and 3) 'a feel good project'. The problem is, 'How do we get the most of this?' The ultimate problem is 'who can speak for somebody else?' (Huntington, 1996).

The Bio-Politics of Knowledge

Among the indigenous peoples there are some concerns over where the use of indigenous peoples' knowledge is leading. Ritva Torikka from the Sámi Council mentions 'the concern that indigenous peoples' knowledge is used as a political weapon, because there are not enough possibilities to cooperate'. However, she does admit that there could be a good opportunity to develop a deeper understanding of the problems related to overgrazing reindeer in northern Finland with the help of indigenous peoples' knowledge (Torikka, 1996). Mary Simon warns that indigenous peoples' knowledge gets 'a lot of lip service' (Simon, 1996).

Governmentality ties subjects to someone else by control and dependence by one's own identity and self-knowledge. Both meanings suggest a form of power which subjugates and makes subject to (Foucault, 1984a, p.21). In becoming and being experts there is a price to pay; it means subjecting oneself to the network of power relations. Strategies of challenging colonial structures also include claims to knowledge. It is possible not to escape power per se but to escape the particular strategy of power relations that directs one's conduct. Each adverse relation is potentially reversible (Simons, 1995, p.85).

This aspect of knowledge emerged in the discussions on the work of the AMAP; how are the indigenous peoples seen as part of the environmental problems and their management in the Arctic. Human health research is a sensitive issue for indigenous peoples. The dilemma is especially difficult in communities where traditional foods are vital to spiritual, cultural and physical well-being. An Inuit view on the problem is: 'We are what we eat. Inuit eat Inuit foods. When I eat Inuit foods, I feel like myself again' (quoted in Huntington, 1997).This is a problem difficult to govern.

The suggestions made at the AEPS seminar in 1994 were that first of all, indigenous peoples should be involved in the scientific monitoring of pollutants and contribute their observations on pollution effects. The practical suggestions included that the AMAP data management program's meta-database-system could be used for collecting information about indigenous knowledge sources and as a registry of experts. Then, the workshop speakers focused on the need for enhancing the possibility for indigenous peoples to contribute their knowledge to the work of the AMAP through a research partnership. Finally, to improve the chances of participation in this kind of work, indigenous peoples needed training and educational opportunities. The seminar report stated that indigenous peoples needed to have better access to monitoring data. Indigenous peoples' organizations have pointed out that it is not enough to only receive information about the results generated by AMAP programs but that they are also given the opportunity to contribute to the design of research undertaken by the AMAP and to point out areas for research based on information from their communities. Representatives from indigenous peoples' organizations have expressed the need for improving the flow of information from the work of the AMAP. They have also emphasized that such information must be presented in a way that is readily understandable for the indigenous communities (Hansen, 1994, p.8-9).

Indigenous peoples' organizations are observers and they have attended the meetings of the AMAP Working Group. The idea was that they would be involved in the planning of the AMAP assessment process, in drafting sections of the assessment and in compiling the information required for the human health and radioactivity assessment work. Representatives of the indigenous groups also participated in the expert groups responsible for both parts of the assessment (AMAP, 1996). In the assessment work of AMAP, the role of the indigenous peoples' community was restricted to presenting two chapters about indigenous peoples and the health issues raised in the AMAP report. In addition, a workshop was held to present papers on indigenous peoples' health issues. Only a handful of the conference participants followed this workshop. The indigenous groups themselves seem to be happy with their role in the AMAP assessment; the 'AMAP had done the Inuit community a great service, and would like to see CAFF incorporate more traditional knowledge in its work and decision-making' (Lynge, 1997).

In practice, the idea of using indigenous peoples' knowledge has been limited. The reason for this, according to Lars-Otto Reiersen, is 'that there was no indigenous peoples' knowledge available to be used'. The ICC suggested a research program to identify such knowledge. In the Reiersen view, this proposal exemplifies the problem: 'there is no such knowledge available since it has to be collected' (Reiersen, 1996).

The politics of governmentality is one of bio-politics; the life of populations is an object of political and scientific concern. State administrators express their concepts of human welfare and state intervention in terms of biological issues. From AMAP's point of view, 'the indigenous peoples are the primary human population at risk from persistent contaminants and other forms of pollution in the Arctic' (AMAP 1993a, p.55-57). The AMAP report says, 'indigenous peoples who rely on traditional diets are likely to be more exposed to several toxic substances than the majority of people elsewhere in the world'. There are no illnesses yet reported in Arctic for which contaminants are known to be a direct cause (AMAP, 1997, p.172).

These practices of surveillance enhance the power of states; the question is how individuals and groups are subjected and categorized by age, ethnic group, sex, disease, and welfare regimes (Dreyfus and Rabinow 1982, p.140). The AMAP dietary assessment is based on toxicological data and biological monitoring. The data is collected by: sampling blood from mothers, from the umbilical cords of the newly born, from placental tissue and by collecting breast milk samples. Questionnaires and the collection of food items can also gather data for monitoring. This kind of research also includes ethical concerns, such as all the participants in human biomonitoring programs being fully informed of the objectives and scope of the programs and having provided written and informed consent prior to their participation (see AMAP, 1995; see also AMAP 1993a, p.55-56). The practice of governmentality is also followed in the Arctic; these political practices include a population surveyed and listed, observed, and examined.

The problem of assessing tolerable risks, according to AMAP, is a challenge: there is 'an urgent need for an in-depth assessment of the toxic effects of all persistent organic contaminants, including the combined effects of these substances'. For example, almost all of the organic pollutants in the study can be detected in breast milk. Sometimes they are found at levels at which the child's short-term exposure is higher than tolerable daily intakes calculated for lifetime adult exposure. Nevertheless, at this time, studies have evaluated the potential effects of persistent organic pollutants in breast milk to 'be limited and not conclusive'. The benefits of breast feeding outweigh the known risks from persistent organic contaminants and breast feeding should continue (AMAP, 1997, p.186).

From the point of view of governmentality, the administrative appara-tuses of the states pose welfare in terms of people's needs and their happiness. The concern for AMAP is 'balancing the information'. It means informing the population about the problem with its negative implications without causing too much panic. This is the responsibility of governments according to AMAP. Giving 'balanced' information about the situation is considered a challenge. First, many factors contribute to health and illness: socioeconomic conditions, health services, societal and cultural factors, individual lifestyles and behavior, and genetics: 'Contaminants enter this already complex scene'. The variation in human exposure depends on varying environmental concentrations of contaminants, local physical and biological pathways, and the local dietary habits of people. Exposure to persistent organic pollutants is the primary concern. Some indigenous groups are exposed to levels that exceed established tolerable intake levels (AMAP, 1997, p.172-173).

Second, fear of contaminants and changes in traditional ways of life can affect both community social structure and individual well-being. The potential negative effects of contaminants to the human health have to be balanced with the positive effects that consuming traditional foods have. The traditional diets are high in animal foods and are rich sources of protein and vitamins. Overall, traditional diets provide 'a strong nutritional base' for the health of Arctic peoples. Market foods from outside the Arctic often have less protein and iron but more fat and carbohydrates. Moreover, changes in food habits follow changes in the way of life leading to a more sedentary life style. Therefore, a move away from traditional foods could contribute to poor health, and in-crease the risk of diabetes and cardiovascular diseases (AMAP, 1997, p.174).

Governmentality makes the population the focus of administering.In governmentality, human needs are now seen instrumentally (Dreyfus and Rabinow, 1982, p.140). The AMAP report aims at developing some scientifi-cally based criteria for the human intake of pollutants. The main criterion for the AMAP report is human tolerance to toxic effects of contaminants. Analyz-ing tolerance is not, however, a simple task. People are exposed to a mixture of many different compounds simultaneously, often at low levels and over their entire lives. The AMAP report talks about a 'tolerable daily intake' which includes 'safety factors' for humans. The safety factor is an attempt to account for the unknown: 'The greater the uncertainty in the toxicology, the larger the safety factor'. When the intake of contaminants in food exceeds tolerable daily intakes, it is a warning that health effects cannot be ruled out and that the situation has to be examined more closely (AMAP, 1997, p.173-174).

For Foucault, this kind of medical discourse is part of the system of administrative and political control of the population. The medical discourse and political practice are connected - not through concepts, methods and utterances of medicine but by political practices that make a possible object for medical discourse (Foucault, 1991c, p.68). The Arctic case is a good example of this. Pollution becomes a problem of population for national governments. The indigenous peoples in the region are recognized as victims of pollution and their contribution is important to the understanding of the problems of the region from a position of victims. In the work of AMAP no signs of accepting this rival knowledge can be found. There are few signs of recognizing their contribution as experts in the local conditions or recognizing the value of different forms of knowledge.

4 Discourse on Development

The Need for a Comprehensive Approach

Pollution as a Common Enemy

The Arctic was mainly perceived as a military theater until the end of the Cold War. This was characterized by the political, economic and strategic division of superpower relations. The Russians were first to see a new identity for the Arctic:

> The Arctic is a region which just a few years ago was seen mainly in the light of military and political interests, a region where many requirements for cooperation in civilian projects were put aside, left of the curb of the 'highway' of international cooperation (Address of the Soviet Representative, 1989, p.1).

According to the Russian representative, 'one cannot say that military and political problems in the Arctic have been resolved, they are very palpable'. However, the situation made new political thinking possible - even in the Arctic. It is 'perestroika that gave the chance to achieve a qualitative change in politics, public attitude in the very atmosphere surrounding environmental protection problems' (Address of the Soviet Representative, 1989, p.1-2).

Transforming the identity of the region dominated by military concerns to one of environmental concerns has not been easy. Some countries, such as Norway and the United States, clung to their old identities. For a long time, the U.S. representatives stressed the strategic importance of the Arctic: 'It [the Arctic] has also been a region of vital security interest to most of our countries' (Weinman, 1991, p.1). It took till 1994, before the United States redefined it Arctic Policy. Then they emphasized cooperation in the region instead of U.S. security interests. The importance of protecting the Arctic environment and conserving its biological resources was initially stressed in the new policy. The need for assuring that natural resource management and economic development in the region were environmentally sustainable was next considered important. Finally, the idea of strengthening institutions for cooperation among the eight Arctic nations was accepted (United States Announces New

Policy for the Arctic Region, 1995).

Studying this discourse shows that the end of the 1980's was a change in the collective identity of the Arctic and the future of the region. Existing identity commitments broke down. This breakdown made an examination of old ideas about self and others possible. It was followed by new structures of interaction. The experience with the AEPS shows that the process so far has been one of continual identity-building for the actors and the region itself as a community. Actors acquire identities - relatively stable, role-specific understanding about self - by participating in the production of collective meanings. Identities are inherently relational. The commitment to, and the salience of, particular identities vary but each identity is an inherently social definition of the actor grounded in the theories which actors collectively hold about themselves and one another. These constitute the structure of the social world. Rethinking those identities paved the way for new identities and practices of interaction. It is not enough to rethink ideas about self and others, since old identities have been sustained by interaction with other actors, the practices of which remain a social fact for the transformative agent. To change oneself it is often necessary to change the identities of those that help sustain the frameworks of interaction. However, interaction may also further change those identities; interaction is a process of defining and redefining identities (see Wendt, 1996). The important point for regime study is, as Stephen Krasner once noted, international regimes 'once established... take on a life of their own and develop their own inner dynamics' (Krasner, 1985, p.77-78).

The Arctic was reframed for the process of rebuilding the identities of the actors. The Arctic had to be given a new identity. Circumpolar countries set themselves a challenge to combat the common enemy - the threats to the fragile environment in the north. The Polar Basin was defined as 'the final depository of a number of air and seaborne pollutants' (Consultative Meeting on the Protection of the Arctic Environment, 1989, p.2). The circumpolar countries developed 'a common understanding of the serious threats to the fragile Arctic environment' (Haarder, 1991, p.2). The main aim of the strategy for Arctic cooperation was to 'identify, reduce and, as a final goal, eliminate pollution' (AEPS, 1991, p.4).

The main concern within the AEPS is pollution control for the sake of human health within the region. Concern for the environment united the actors and made the AEPS possible for a short period. Concern for the environment, in particular the concern over pollution, was the lowest common denominator for cooperation in the Arctic (Prokosch, 1996). Politically, the issue of pollution lost its importance within the framework of the AEPS. At the 1996 Inuvik meeting, the issue of pollution is not among the first concerns mentioned; it was considered a problem for the 'long-term health of Arctic ecosystems' (Inuvik Declaration on Environmental Protection and Sustainable Development in the Arctic, 1996). Politically, the ministers at Alta in 1997 concluded, after receiving the AMAP report that:

> The Arctic, in comparison with most other areas of the world, remains a clean environment with large areas of unspoiled nature (Alta Declaration on the Arctic Environmental Protection Strategy, 1997).

The ministers noted that 'in some parts of the Arctic severe pollution from local sources requires both national and international remedial action' (Alta Declaration on the Arctic Environmental Protection Strategy, 1997). However, the sense of urgency in dealing with Arctic environmental concerns and the sense of the fragility and vulnerability of the Arctic environment was lost.

A Rich Resource Region

The Arctic was seen more in terms of its rich natural resources and their potential for further regional use and development, instead of the fragility and vulnerability of its ecosystems and populations. The Canadian representative stressed the identity of the Canadian Arctic as '...a land surprising rich with flora and fauna and it offers enormous resource potential' (Campeau, 1990, p.1). According to the Russian representative:

> For us the Russian Arctic is not only a vast territory, where about 2 million Russians live, not only a considerable part of our economical potential and not only a unique component of our nature. The Arctic for us, especially now is something much more (Danilov-Danilian, 1993, p.1).

The Arctic provides the Russians with: '... the ores of Murmansk district, nickel and copper of Monchegorsk and Pechenga region, bauxites of Archangel District, coal of the Pechora Basin, oil and gas of the Timan-Pechora gas province'. Major mineral extraction and refining complexes at the country's largest metal processing plant are located at Norilsk in Arctic Russia. The 'next step will be the development of oil and gas fields on the Arctic seashelf,' according to the Russian representative (Address of the Soviet Representative, 1989, p.3).

Fisheries and hunting are of special importance for Greenland. In terms of non-renewable resources:

> ... granted mineral and hydro-carbon wealth should be exploited. Nobody can

expect us to have this kind of development potential just lying around untouched (Statement of the Danish Delegation, 1989, p.2).

For the Danish, 'natural resources in the long run can only benefit the people if they are utilized in a sustainable way'. Denmark is 'no friend of the 'boom and bust' mineral exploitation that has caused so much havoc in the Arctic'. The Danish also noted that the 'exploitation of mineral resources must take place within the framework of a general development policy that takes into account a number of other important requirements of a social cultural and especially environmental characters' (Statement by the Danish Delegation, 1989, p.2).

The Norwegian representative described the resources the high Arctic gave Norway: fisheries, petroleum related activities, industrial activities and efforts to use the resources allowing for environmental considerations. Arctic natural resources are expected to receive more attention in the future because 'not only has the exploration of potential resources in the Arctic acquired new importance' but new technologies allow for their utilization (Statement of the Norwegian Delegation, 1989, p.1).

The issue of development and development opportunities became the focus of attention. As Finland said: 'the direct and indirect effects of modern, developed models of production and life in the northern hemisphere call for immediate attention' (Bärlund, 1989, p.1).

Equitable Development

In 1990, the basis for the involvement of the indigenous peoples was that 'indigenous peoples have a great deal at stake'; the culturally-distinct societies of indigenous peoples are 'vulnerable to the effects of environmentally-unsound or other inappropriate development activities' (Simon, 1990, p.2-3). The representatives of ICC challenged the 1990 draft for the AEPS. According to the ICC the strategy should have several diverse objectives instead of one, which concentrated on the need of protecting the Arctic environment. These objectives relate to the 'northern peoples, as well as Arctic ecosystems'. The strategy also had to be sustainable and equitable in ecological, social, cultural and economic terms. The Inuits required that the strategy and its objectives were put into the larger perspective of 'development, peace and human rights' (ICC, 1990, p.1).

The Inuit particularly focused on the issue of 'equitable development,' by demanding that this should enhance the development of indigenous peoples and respect their rights. From an indigenous perspective, development projects within the Arctic have neither been sustainable nor equitable. The adverse impacts from these activities, both within and outside of the Arctic, are directly borne by indigenous peoples and by the living resources upon which they depend. In particular, if the injustices and imbalances of current development practices are not effectively addressed, their health and traditional ways of life may be irreversibly affected (Simon, 1990, p.2). The issue is also one of the right to use living resources in the Arctic: 'within the limits of sustainability, all indigenous peoples of the Arctic must be able to exercise their right to utilize living resources' (Hansen, 1994, p.16).

The Inuit emphasize the right of participation in the management and development of the Arctic and its resources. From the indigenous perspective, development including or affecting indigenous territories must not undermine, but rather enhance the economic, social, cultural and political development of indigenous societies. Development in or affecting indigenous territories must fully respect the rights of indigenous peoples and accommodate indigenous values and concerns. Development must not be imposed on indigenous peoples without their free and informed consent. Development initiatives by indigenous peoples should be encouraged, by ensuring significant and accessible opportunities that include government assistance and support. Development should only take place at a rate and pace compatible with the local communities affected. Indigenous peoples must participate equitably in the benefits of development, in a manner acceptable to them. Finally, development policies and objectives must appropriately recognize the importance of the environment and its resources in the survival and growth of distinct indigenous societies. These concerns should be taken into account by the ICC in planning Arctic environmental cooperation (Simon, 1990, p.2-3).

The Sámi understanding of sustainable development was based 'nature's own premisses'. Utilization must represent 'a balance between what nature can give and what we take from it,' so that nature will not be depleted. Economic considerations should not govern utilization solely. The aim should be to minimize the production and use of products which deplete natural resources (Sámi Programme on the Environment, 1990, p.5). In the Sámi view, they are part of the ecosystem. Their 'cultural manifestations are adapted to an ecological balance between what nature can give and what we can utilize in relations to the nature's productive capacity' Also, Sámi cultural tradition teaches the Sámi 'how nature is to be used without being consumed'.The Sámi representative noted that the Arctic cultures (based on e.g., fishing, hunting and reindeer herding) have successfully applied the principles of the sustainable development strategies for centuries and so they can serve as 'very valuable models' for these strategies in the modern technological world (Aikio, 1990, p.6). Their culture is a living culture which enables them 'to adapt to various natural conditions, acquiring new knowledge which will enable us to survive' (Sámi Programme on the Environment, 1990, p.3). For the Sámi, however, the emphasis is on the right of development. According to this view, the Sámi have:

... inalienable right to preserve and develop our economic activities and our communities in keeping with our common conditions and together we wish to preserve our lands, natural assets, and national heritage for future generations (Halonen, 1991, p.1).

According to the Sámi, when land and water resources are exploited by people other than the Sámi and for purposes other than those of the Sámi, the Sámi must be given a share in the profits of exploitation. The Sámi demand that primary occupation must be protected and promoted on their terms and reindeer herding must be reserved for them by law, providing strong rights of access to grazing resources (Sámi Programme on the Environment, 1990, p.6).

The Sámi see the problem as being 'subjected to constant influences which is transforming our pattern of life and our relationship with nature' (Sámi Programme on the Environment, 1990, p.2). According to the Sámi view, 'the Finnish initiative for the protection of the Arctic has resulted in further cooperation among the Arctic governments. This we believe is of great significance of the existence of the indigenous cultures. It is important, in this regard, to consider whether the agenda for the next ministerial conference would include the issues regarding Arctic indigenous peoples' (Halonen, 1991, p.3).

The Russian indigenous peoples consider it essential for indigenous peoples to be guaranteed the possibility of continuing their traditional ways of life. As the chairperson of the Association of Indigenous Minorities of the North, Siberia and Far East of the Russian Federation, V. Sanghi states the 'fundamental' requirement for saving the indigenous peoples in the Russian North is to respect the lifestyle of the peoples: 'The lifestyle is what makes them who they are' (In, Alia, 1991, p.28). One of the main prerequisites for the successful development of the peoples of the north is that they are granted the opportunity to plan their own present and future. The aim of the Association is to involve its members in activities related to monitoring the ecological situation and the preservation of flora and fauna in the territories of indigenous peoples (Indigenous Peoples of the Soviet North, 1991, p.54-55).

Together the indigenous peoples stress that they 'desire not only to survive but to thrive as Indigenous Peoples into the 21st Century'. They require 'sustainable and equitable development' in their homelands. This kind of development includes subsistence hunting and renewable resource harvesting. For them, the exclusive and collective right to lands and resources for subsistence, and involvement in all decision-making processes concerning the management, research, and allocation of resources, is important. Indigenous peoples do not want to be left on the 'outside' of development, as has often been pointed out by both the representatives of the ICC and the Sámi Council in their statements. They do not want their regions to be developed 'over them'; they want to have an input in those developments.

In particular, according to the Arctic Leaders' Summit, there 'is a need to find an appropriate balance' between the ongoing subsistence activities of the indigenous peoples and the development of renewable and non-renewable resources by applying the principles of 'sustainable and equitable development'. The Arctic Leaders' Summit declaration criticizes 'the antiharvesting lobby movement' which, according to the declaration, 'misrepresent to the public and governments their objective which is the complete prohibition in killing wild animals'. The antiharvesting movement causes 'great harm to indigenous peoples and indeed, places in jeopardy their very right to exist as distinct peoples' (Faegteborg, 1993, p.39).

Redirecting the AEPS

In 1991, the Arctic states agreed that the AEPS should 'allow for sustainable economic development in the North so that such developments does not have unacceptable ecological or cultural impacts' (AEPS, 1991, p.2). It was, however, noted in the preparatory process leading to the Nuuk ministerial meeting that 'the AEPS presently projects a very protectionist posture concerning the Arctic despite the many references within the Strategy to sustainable development' (Protecting the Arctic Environment, 1992, p.5).

The protectionist image of the AEPS was eroded in the Nuuk and Inuvik Declarations. The name of the 1991 declaration, the 'Declaration on the Protection of the Arctic Environment' was changed in 1993 to 'The Nuuk Declaration on Environment and Development in the Arctic'. The countries decided at the Nuuk conference in Greenland (1993), to conserve and protect the Arctic environment for the benefit of 'present and future generations' and for the global environment. The understanding at the Nuuk meeting was for sustainable development, which meant that environmental protection constituted, 'an integral part of the development process and cannot be considered in isolation from it' (The Nuuk Declaration on Environment and Development in the Arctic, 1993).

Arctic environmental cooperation became a part of the Rio rhetoric in 1993. In the Declaration of the Second Ministerial Meeting on Arctic Environment in Nuuk, ministers stated, 'we support the achievements of the United Nations Conference on Environment and Development, and state our belief that the principles of the Rio Declaration on Environment and Development have particular relevance with respect to sustainable development in the Arctic' (TFSDU, 1994, p.2). By 1993, there was a shift in defining the source and the nature of the threat to the Arctic environment; the issue of the use of local natural resources captured attention - at the expense of the pollution problem. This concern included the utilization of natural resources by indigenous peoples.

As a sign of this commitment the Arctic states established a task force on sustainable development (Task Force on Sustainable Development and Utilization, TFSDU). Terry Fenge from the Canadian Arctic Resources Committee (CARC) described the establishment of the task force as, 'an interesting departure' broadening the AEPS (Fenge, 1995). The status of the task force was upgraded to that of a working group, reflecting the commitment of the states to the continuation of work on sustainable development.

The report on the work by the TFSDU in 1995, concludes that the competing approaches to sustainable development include some common conclusions. First, they suggest that 'long-term social well-being, economic development and environmental health are interdependent'. Second, what is needed is 'the full integration of environmental costs and benefits into accounting and policy assessment procedures in order to inform economic decision-makers more completely about the trade-offs required'. Finally, 'it is crucial that those affected by economic development should be involved in decision-making around new plans and initiatives, in order to ensure that he values inherent in their culture are reflected'. The human-centered and culturally-oriented approach to the human-environment relationship in these formulations is clear; the cultural factor is 'a factor of critical importance in the Arctic' (TFSDU, 1996, p.16).

However, the concept of sustainable development seemed to be left more or less open. The main message of the AEPS task force on sustainable development and utilization seems 'to avoid superficial, theoretical work' (Hurst, 1994, p. 125). It almost seems that despite the lack of interest in the 'theoretical' discussion on sustainable development, the AEPS participants are reinventing the whole idea and are swallowed up by procedural discussions. Even should the participants try to avoid theoretical thinking about sustainable development there is some thought about it in the future practices of cooperation developed by the working group.

The TFSDU was established in 1993 to:

... propose steps governments should take to meet their commitment to sustainable development in the Arctic, including the sustainable use of renewable resources by indigenous peoples, taking into account that management, planning and development activities shall provide for the conservation, sustainable use and protection of Arctic flora and fauna for the benefit and enjoyment of present and future generations, including local populations and indigenous peoples (Hurst, 1994, p.123).

The task force was expected from preparing reports and making recommendations to the ministers on the following: 1) identification of the goals and principles of sustainable development in an Arctic environmental contexts and finding opportunities and mechanisms for the application of these principles; 2) opportunities to enhance indigenous peoples' economies, and to improve the environmental, economic and social conditions of Arctic communities through the sustainable utilization of natural resources, while protecting the cultures of indigenous peoples; 3) specific issues and problems presented to the conservation, sustainable use and protection of Arctic flora and fauna by management, planning and development activities, and proposals for measures to mitigate or resolve such issues and problems; and 4) the needs for new knowledge, ways of facilitating communication and the sharing of information concerning the application of new or proven technologies and management practices (TFSDU, 1994, p.4; Hurst, 1996; Snider, 1996).

Under the TFSDU, the idea of an Arctic Agenda 21 was developed. Six themes were examined more closely in the regional Agenda 21: circumpolar cooperation, poverty, decision-making, the conservation of biological diversity, the protection of the oceans and environmental threats. The theme of circumpolar cooperation focuses on the economic development of the Arctic region and the problems caused to it by restrictions on the import of wildlife products. Circumpolar cooperation aims at the development of international trade and environmental policies to support the local economies. Combating poverty aims to provide all persons in the Arctic with: the opportunity to earn a sustainable livelihood and to develop integrated strategies; programs of sound and sustainable management of the environment, resource mobilization, poverty eradication, and alleviation; employment and income generation and education and training. The chapter on integrating environment and development in decision-making aims at including environmental and developmental considerations into policy, planning and decision-making at all management levels in all sectors of circumpolar society (TFSDU, 1995b, p.7-13).

The three other topics; the protection of the oceans, the protection of biological diversity and environmental threats overlap with the activities of other working groups. It reflected the aim of developing the task force to a higher level compared with others, so that the task force would have a comprehensive role in advancing sustainable development in the Arctic. The division of duties between the working groups and their relations to each other became an issue when the establishment of the Arctic Council was negotiated. This made the work of the TFSDU difficult in the end.

The Human-Environment Relationship

Ethics and Environmental Discourse

Regimes institutionalize a particular ethical stand on the human-environment relationship. In the anthropocentric (or human-centered) approach, environmental problems are a destructive factor in human life; they limit the possibilities to enjoy healthy, safe and comfortable surroundings. Human-centered ethics emphasize future generations and their possibility to use resources for their own development. Human needs and interests form the basis of the entire ethical system of standards and rules governing the relationship between humans and nature (Passmore, 1974, p.73-81; Eckersley, 1992, p.33-47).

In the biocentric (or life-centered) view, the question relates to factors threatening life on earth. This view is that expansionist human action threatens life-supporting systems. Life-centered approaches do not consider the duties towards nature to be derived from people's duties towards each other. In the life-centered approach, obligations and responsibilities are seen as arising from certain moral relations between human beings and nature. The natural world, in this approach, is not seen simply as an object to be exploited; animals, plants etc., are regarded as more than resources for human use. Living things are considered to have a worth of their own, irrespective of their actual or possible usefulness to human beings (Rolston, 1988, p.216-220; Taylor, 1986, p.12-13).

The concept of sustainable development reflects a particular way of understanding the human relationships between how economic activity and natural resources are arranged. These could be called 'world views' including the aspects of an ethical stance on the relations between man, the environment and the world. According to the World Commission Report on Environment and Development, sustainable development is 'development which meets the needs of the present without compromising the ability of future generations to meet their own needs' (World Commission on Development and Environment, 1987, p.8).

The idea of sustainable development represents the tradition of resource management in defining the human-environment relationship. Resource management is anthropocentric at its core; hurting nature is beginning to hurt economic man. Thus, the instrumental economic paradigm prevails, although it is enlarged to encompass some basic ecological principles in an attempt to maintain ecosystem stability for the support of sustainable development; in short ecology is being economized (see Colby, 1992).

The World Commission demonstrated how economic growth depended heavily on the increasing use of environmental resources and also documented the impact of policies. Many of its conclusions, however, reaffirmed the fundamental premises of conventional development thinking, in particular the stress on economic growth above all else. The main benefit of the concept of sustainable development is that although it still considers economic growth as the main aim, it also includes the ecologic constraint, or 'ecological imperative,' that there are limits that should not be broken. The question of what the limits are and how they should be defined is essentially left open (Turner, 1988, p.5).

Defining the limits of development means defining the basis for the criteria of sustainability. The alternatives are to either use ecological or social criteria. The idea of social development requires that state governments recognize and accommodate the rights of indigenous peoples to self-government, lands and renewable and non-renewable resources. This issue is fundamentally one of recognition of their cultural, social and economic needs. The ecological aspect of sustainable development refers to a management policy which is based on sound ecological principles. These are based on the idea that man is a part of nature and say that renewable resources must be managed. They must be protected in a manner that maintains essential ecological systems and processes and ensures sustainable use of species and ecosystem both now and in the future (Oreskov and Seijersen, 1993, p.12).

Interpretations of sustainability were many in the Arctic cooperation. The Swedish delegation noted that 'today's system of environmental protection in the Arctic region was fragmented'. 'An overarching ecological perspective could provide a more comprehensive assessment of the existing situation' (Statement by the Swedish Delegation, 1989, p.1). According to the Swedes, this approach

> ... must include provisions for the conservation and management of living as well as non-living natural resources. With regard to living resources, a system that guarantees the long-term sustainable utilization of these resources and prevents the over-exploitation of species and populations is urgently needed.

Such a system must also contain regulations that ensure the recovery of species and populations which have been over-exploited in the past (Dahl, 1991, p.6).

In their view sustainable development is 'a matter of mutual dependence rather than control and exploitation'. This means that nature as an element has to enter into all different aspects of the decision-making process. Sustainability means more efficient and environmentally sound uses of natural resources and the introduction of ecologically sound technologies. It requires that environmental stress and environmental emergencies are prevented in beforehand. The Swedish saw the cooperation as including an 'Arctic Sustainable Development Strategy' which could express objectives and principles on how to protect and sustainably utilize the Arctic environment (Introductory Statement by Assistant Under-Secretary, Desiree Edmar, 1990, p.2-3).

The critical question for the Norwegians is to find 'the right balance' between resource conservation and resource utilization. Not least is the question of securing the long term basis of livelihoods for the local people (Statement of the Norwegian Delegation, 1989, p.11). Several principles were suggested for dealing with the challenge of balancing the use and protection of natural resources. In the Norwegian view:

> Only through careful stewardship and with due regard to the precautionary principle will we be able to prevent further environmental damage and degradation and thus ensure rational use of the regions resources, to the benefit of its peoples and mankind as whole (Stoltenberg, 1991, p.5).

They proposed a broad ecological context to deal with the challenges. This broad approach would cover interrelations between different species and the environment as a whole, which must be protected from pollution and degradation. They stated: 'to secure sustainable development in the Arctic is a challenge which is part of the broader global challenge to safeguard our common future' (Statement of the Norwegian Delegation, 1989, p.16).

The problem of sustainability, for the Russians, was a problem of including the environment within economic considerations. The major problem was: 'imperfection of the economic mechanisms that do not account for ecological factors'. The production dominated, sectoral approach to the utilization of Arctic natural resources with an imperfect management system for the utilization of the natural environmental added to the problem. The solution in the Russian view was equipment and technology for use in the Arctic environment, and knowledge about the environment. Finally, what was needed was an 'environmental utilization strategy' including a 'clear concept of economic development in the Arctic for the benefit of both economic and ecological interests' (Address of the Soviet Representative, 1989, p.5).

The ecosystem approach, according to the Canadians, linked to the concept of sustainable development and based on sound scientific evidence, is the most suitable for studying and managing the Arctic environment. Their view on the ecosystem approach was that it enabled everybody to 'ensure the future health and well-being of Arctic ecosystems, thereby enabling countries to fulfil their natural and international responsibilities' (Proposal by the Government of Canada, 1990, p.2).

In the ecosystem approach 'nature and human development are dynamic by definition'. It focuses on essential ecological components, systems and biodiversity and minimizes climatic change; contaminants are seen as a complex problem in the ecosystem. The ecosystem model divides sources as local and distant; transport is atmospheric, fluvial, in ocean currents and fresh water. Impact is divided into four categories: terrestrial, marine, exposure/diet and human health. The basic principles of the approach are: each segment of the ecosystem model links to all of the others, policy choices can be drawn from an understanding of how relevant components of the ecosystem interact with each other within the context of the dynamics of contaminants transfer, and the focus of resources and science on only one sector at cost of another will not provide the basis for policies (Final Draft WG1 Arctic Environmental Protection Strategy, 1991, p. 4-5).

Man is an 'inherent part of this ecosystem'. The human dimension has particular emphasis in the ecosystem approach, and the effects of environmental degradation to health and the northern way of life (these are heavily reliant on Arctic ecosystems) are highlighted (Statement by the Canadian Delegation, 1989, p.3-4). The human dimension is, 'the incorporation of the social, economic and cultural needs and values of Arctic peoples through the development and implementation of the strategy' (Proposal by the Government of Canada, 1990, p.2). It was not until 1995 that the Canadians announced they had dropped the idea of developing an international application of their ecosystem approach due to financial limitations (see TFSDU, 1995a).

Sustainable development is not only a theoretical concept, it is connected with practices. Thought, such as the concept of sustainable development, is not something superfluous for Foucault, a superstructural reflection of social reality. Thought is an integral part of action. When we think, speak and behave we do so in relation to thought. Thought is not independent of, but is related to, economic, social and political determinations (Foucault, 1984b, p.334-335). Practices rest on 'modes of thought' where 'thought is... often hidden, but... always animates everyday behavior. There is always a little thought' (Even in the most stupid institutions, Foucault adds). There is thought in 'silent habits' (Foucault, 1988a, p.155).

Foucault directs attention to political practices: to those practices that make the objects of discourse, and to the practices that make the discourse possible - the political practices that define what is 'sayable' and 'doable'.Thought is materialized in the discourse. Discourse also has material effects on practice. Therefore, the dependence of thought on practice must be also recognized.

Conservation and Biodiversity

Few ideas of conserving nature for its own the sake were presented. In 1989, the background paper considered the conservation needs of the Arctic:

> ... the unique character of the Arctic region as one of the few remaining wilderness areas on the earth [and] it would be important to proceed much further than traditional fora, fauna and habitat protection and management (Jaakkola, 1989, p. 5).

The more conservationist view was that it would have been of greater value if the Arctic countries could have agreed on a 'conservation policy' for the Arctic region and prepared a strategy for its implementation. The background paper suggested that, rather than start a completely new process, it would be more feasible to utilize the already ongoing elaboration of the existing international arrangements (Jaakkola, 1989, p.5).

For the more conservation oriented AEPS parties the Arctic is a region where the environment needs to be protected, not only used for resources. After the proposed areas for protection listed by the CAFF are established about 16% of the Arctic land area was protected. These areas are not fully representative, since some 'key habitats' - such as isolated islands, fjords, coastal areas, marine areas and wetlands - are under represented. Even with the new proposals being implemented the situation will not change. The network will not represent 'the variety of the Arctic ecosystems' nor will it 'contribute to the maintenance of viable populations of all Arctic species'. The need for planning coastal and marine protected areas is stressed in the CAFF report. Some 100 of the present 285 protected areas have a marine component. At the present there are five transfrontier conservation areas in the Arctic (CAFF, 1997a).

CAFF notes that maintaining the biological diversity of species and populations is considered 'fundamental' to the health of Arctic ecosystems, to the global biosphere, and to the continued welfare of indigenous peoples of the region. Biodiversity refers to:

> ... the variability among living organisms from all sources including, inter alia, terrestrial, marine and other aquatic ecosystems and the ecological complexes of which they are part; this includes diversity within species, between species and of ecosystems.

This is the definition of the Convention on Biological Diversity signed at the UNCED in 1992 (CAFF, 1996, p.5).

The Arctic ecosystems are considered 'comparatively simple in a global context,' however, 'they are unique in terms of their biological diversity and because anthropogenic impacts are still at a relatively low level compared to other parts of the earth'. The characteristics of Arctic ecosystems and their biodiversity also make them more vulnerable to human impacts and susceptible to threats. Arctic species have adapted to the harsh environment which may make many of them unique in genetic variation (CAFF, 1995-1996, p.7).

According to CAFF's list of threats to the Arctic, 'the ecosystems and the biodiversity of the Arctic can be threatened by many human activities and the Arctic countries have identified several of particular concern in their respective countries of the region as a whole'. Major threats to Arctic habitats are primarily related to activities associated with the exploitation of natural resources, the development of an infrastructure, the increased public uses of natural resources, and pollution (CAFF, 1994, p.87).

The report considers at length the threats related to mining, hydroelectric power development and oil and gas exploration. Badly managed offshore mining, for example, may affect seabird feeding grounds or cause pollution from metals found in sediment, such as mercury, arsenic, cadmium and lead. Building oil fields may result in direct habitat loss due to road building pads, and indirect loss due to alteration in drainage patterns, dust deposition and contaminants. The landscape is continuously altered by a network of roads, pipelines and facilities. The presence of these facilities can influence movement of animals, such as caribou and waterfowl (CAFF, 1994, p.88).

The continuing exploration, production and transportation of oil pose the threat of a major oil spill in an area of very high biological productivity. A spill of any size could have 'a significant negative impact'. Spills of crude oil, refined fuels from ships and from onshore and offshore oil facilities are 'significant threats' to marine wildlife. The development of northern industries and communities will affect the natural environment (CAFF, 1994, p.88).

The expansion of modern forestry in the Arctic has drastically increased the threat to forest species of plants and animals. The fragmentation of habitats can lead to 'species loss and endangerment of and reduction of genetic diversity'. Overgrazing by reindeer is an increasing problem both in all Scandinavian countries and in Russia. Overfishing threatens marine food chains and populations of sea-birds. The cumulative impact of the tourism industry, including a formal and informal transportation infrastructure, waste and noise and increased population, could be 'substantial' (CAFF, 1994, p.89-90).

The CAFF report concludes that 'there are many categories of threats to terrestrial, limnological and marine habitats in the Arctic, some actual and some potential'. These threats are directly and indirectly related to the consequences of human activities. Most of them arise from local and regional activities. 'All these threats constitute a great challenge to the management of Arctic nature and its resources' (CAFF, 1994, p.93). In 1995, CAFF defined the 'priority threats' to Arctic biodiversity as:

> Group 1: mineral and petroleum exploration and development, road infrastructure, habitat fragmentation, motorized vehicles;
> Group 2: rapid expansion of tourism;
> Group 3: fisheries practices and bycatch, overexploitation of species/hunting pressures, oil spills; and all pervasive: climate change (CAFF, 1997b).

In emphasizing the need for protection another concept, biodiversity, could have become a principal for cooperation instead of sustainable development. The idea of biodiversity does not necessarily exclude indigenous peoples and their concerns. From CAFF's perspective, indigenous peoples are:

> ... people who have lived in the northernmost regions of the Arctic for about 100 000 years. Over time, they have developed ways to survive in the harsh climate of the Arctic and to use biological resources in a sustainable manner thus creating a long-lasting and close connection between the people and the ecosystem (CAFF, 1996, p.22).

The draft emphasizes the value of traditional and local communities in the conservation and sustainable use of biodiversity: the 'preservation and use of traditional knowledge that has resulted from this long-term interaction (indigenous peoples') with nature has been extremely valuable in the conservation and sustainable use of Arctic biodiversity' (CAFF, 1996, p.22).

'Conservation issues do not enjoy the same level of recognition and acceptance within the AEPS' (Report of the CAFF Chair and Secretariat, 1996, p.1). Experience so far shows that the dominant priorities for the AEPS are sustainable economic development and pollution control. The idea of advancing biodiversity in the Arctic was included in the final AEPS declaration in Alta in 1997. Here the states recognized 'the importance of biological diversity in the Arctic region' (Alta Declaration on the Arctic Environmental Protection Strategy, 1997). However, this idea will mainly be used mainly as a framework for future CAFF activities. Having a biodiversity strategy in the Arctic is not an uncontroversial matter; the U.S. government has had some reservations about it both in the Arctic and in the global context (Kankaanpää, 1996).

Environmental Security

During Arctic Council negotiations a strong resistance from the U.S. and most other governments led Ottawa not to push military security onto the agenda (Scrivener, 1996, p.22). In the end, all military security issues were excluded from the work of the Arctic Council (Declaration on the Establishment of the Arctic Council, 1996). The connection, however, between environmental concern and security thinking was made in the negotiations. The aim of the Arctic Council is to 'enhance the collective environmental security of Arctic states and peoples, inter alia, through protection, preservation and restoration of environmental quality' (Arctic Council Declaration Draft, August 16, 1995).

It has been outside the negotiations that a greater enthusiasm around the concept has been found. According to the Arctic Committee of Parliamentarians collective environmental security means to 'protect and defend the Arctic against environmental threats arising from outside the region and from unsustainable activities within the Arctic' (Second Conference of Parliamentarians of the Arctic Region Conference Statement, 1996). According to Guy Lindström, the concept reflects the notion that all activities include environmental risks (Lindström, 1996). For the World Wide Fund for Nature (WWF), collective environmental security can be achieved 'through cooperation to conserve, protect, enhance and restore environmental quality' (WWF, 1995, p.1).

The main argument against using the concept of collective environmental security in the Arctic is that it would direct cooperation onto wrong track (Griffiths, 1992; see also Deudney, 1990; Brock, 1991; Dalby, 1992). The critical points on environmental security include the word 'security' as having a strong nationalistic and militaristic burden that cannot be changed just by adding 'environmental' or 'collective' at its beginning. It has been pointed out that defining environmental concerns as security risks is a security risk itself. Developing ideas and practices that interpret environmental issues as security risks and threats are destructive to the development made so far in the Arctic. Adopting the concept and building the circumpolar region according to the idea of collective environmental security would undermine the development of the cooperation of the region so far; it defies the idea that the Arctic could be seen as something other than a military theater. Modernity invaded the Arctic in a special way, having originally arrived primarily and almost exclusively, in a military mode. The predominance of military concerns has suppressed other forms of modernity (Osherenko and Young, 1989, p.122).

Cooperation on military issues takes place outside the AEPS and the Arctic Council at the moment; it is now restricted to the Arctic Military Environmental Cooperation (AMEC) Forum between the U.S., Norway and Russia. According to information, other countries have expressed interest in AMEC but there are no formal plans or agreements to expand the cooperation at this time. Other countries may take part in the individual AMEC projects after agreement among the three partners. The AMEC vision is the 'preservation and sustainable military use of the Arctic environment' (AMEC Information, 1997; see also Kirk, 1997).

Governmentality focuses on the study of ethics, power and politics. The importance of ethical arguments in states' discourse is not always understood: 'it is remarkable how moralistic governments often are in discussing their obligations and those of others' (Keohane, 1984, p.126). For environmental matters, such interest in ethical arguments is anything but strange. The human-environment relationship is based on values. Acknowledging the political nature of these arguments is, however, sometimes difficult for political scientists. However, 'ethos is fundamentally a political question' (Dillon, 1995, p.349). In E.H. Carr's view:

> It is impossible to build up effective institutions, national or international, unless they rest on the function of an accepted common morality: without that foundation no institution can work (Carr, 1949, p.69).

The approach to the environment by mainstream regime theorists follows the tradition of western science. This tradition requires the separation of values and facts and the separation of the subject of knowledge and the object of knowledge. Being rational and objective has required these practices However, facts and values cannot be separated; the human-environment relationship is based on values. Descriptions of the environment and environmental problems include normative evaluations. Those often 'hidden' or 'silent' norms may include assumptions of the human-environment relationships (see Merchant, 1980, p.4).

Understanding cognitive dimensions of regime building are important since regimes are fundamentally cognitive entities that do not exist independently of the actors' ideas on how the world works (see Kratochwil and Ruggie, 1986, p.764). In regime theory, the so called 'cognitivist' approaches on individual representations and perceptions oppose the constructivists' idea of the social construction of problems based on collective significations (Jönsson, 1982; Bonham et al., 1987; Shapiro et al., 1988; Jönsson, 1993). As Emile Durkheim suggests these collective significations:

> ... could not be confused with biological phenomena, since they consist of representations and actions; nor with psychological phenomena, which exist only in the individual consciousness and through it. They constitute, thus a new variety of phenomena; and it is to them that the term 'social' ought to be applied (quoted in Järvikoski, 1996, p.79).

The human being appears in a dense web of language and meanings; everything surrounding the human being has meaning and everything he arranges around him constitutes a coherent whole and a system of signs. Because of language, a whole symbolic universe can be created within which man 'has a relation to his past, to things, to other men and from which he can build something like a body of knowledge, in particular knowledge of himself' (Foucault, 1970, p.351; 357). In this universe, nature does not speak but human beings speak for it. In that speech, not only is the environment constructed but the human being himself is constructed, too (Foucault, 1970, p.160).

Intersubjectivity should be understood as a web of socially constructed meanings. Emile Durkheim's social is not 'collective' consciousness but intersubjectivity - a living relationship and tension among individuals (Harris, 1988, p.12). From a constructivist point of view, the relationship between man, the environment and institutions is reformulated and reconstituted (Berger and Luckmann, 1966, p.168).

Through this move an effort is made to understand the subjective dynamics of interdependence. Interdependence is, above all, a matter of how the world is understood (Rosenau, 1990, p.424, Hettne 1994; see also Sprout and Sprout 1971, 192; Wilde 1991, p.216). The question of interdependence transforms from a question of the relationship between two systems - human and environment - to one of understanding of meanings attached to different interdependencies and the ways people interpret human-environment relationships (see Douglas and Wildavsky, 1983, p.4-5).

In the arctic cooperation, this understanding of interdependence was based on the idea of sustainable development. The Russian representative stated,

> We speak here about the interdependence, which by its nature requires international efforts, joint action of the Arctic nations for the purpose of sustainable development (Danilov-Danilian,1993, p.2).

The hegemonic silence about military activities and their impact on the Arctic environment is breaking. There is no point in endangering the results of initiatives such as AMEC by emphasizing the security related concerns of the states in the Arctic in relation to the environment, not at least in the name of environmental security. The ultimate mandate of the council is to make the area into 'a domain of enhanced civility' (Arctic Council Panel, 1991, p.2-3). The success of the Arctic cooperation is in turning the region into a 'domain of enhanced civility' using the frase of the Arctic Council Panel.

The establishment of the Arctic Council institutionalizes a 'new' identity as a modern region. What is important in the emerging Arctic order of sustainability is that it was not defined in terms of the environmental impact of militarization in the region, but in terms of economic development and development opportunities. Within a regime a hegemonical discourse can be detected. In the Arctic such a hegemonic discourse is one of development. Hegemonical discourses are considered 'natural and taken for granted;' a hegemonical discourse is seen without alternatives. Hegemonical discourse can be found in a regime: 'A regime is the product of a genuinely social process, in which human actors establish - or impose - frameworks of meaning which identify regime related action as acceptable or inacceptable' (Behnke, 1995, p.182).The concept of sustainable development itself and its spread to cover the whole globe as the dominant perspective in defining the human-environment relationship could be defined hegemonical. The Arctic region was normalized to make it one of sustainable development and part of the UN led welfare internationalism (see Suganami, 1989, p.109).

The governmentality in the Arctic is one of normalization of the region. The environmental threat was normalized through cooperation. Normalization is both individualizing and totalizing; it is about finding answers to the question of what it is for an individual, society, or population to be governed or be governable. Normalization is a discursive morality forming the ethical foundation of cooperation (Gordon, 1991, p.36). By normalization Foucault means a system of individuals that can be distributed around a norm - a norm which organizes and is the result of controlled distribution. Both indigenous peoples and states in the Arctic have participated in this process. Normalization is a discursive morality forming the ethical foundation of cooperation.

Sustainable Development in the North

'Two Pillars'

In the negotiations to establish the Arctic Council, the issue of sustainable development emerged as one of the most important items on the agenda. In particular, the question of whether the working group on sustainable development should have special status in respect to the other working groups has often been discussed. The progress made during the summer 1995 was lost at the negotiations in Washington, D.C. in September 1995; one of the most important items in the reopening of the negotiations was the issue of the relationship between sustainable development and environmental protection in the Arctic and the Council's role in promoting them.

The problem of balancing the need of Arctic environment protecting with the use of local natural resources was an important part in the negotiations to establish the Arctic Council. In the original concept, one of the most important substantive issues to be handled in the Council was considered the protection of the Arctic environment (Arctic Council Panel, 1991, p.6). Some commentators have said that giving working group status to the task force on sustainable development was 'in some ways a victory for environmentalism,' making the survival of the 'pure environmental agenda for the Arctic' possible (Scrivener, 1997, p.5). Another critical view saw that 'the AEPS is lost by refocusing the cooperation through the establishment of the Arctic Council' (Brelsford, 1996a). The statement of WWF referred to the responsibility to protect environmental integrity of the Arctic 'for its intrinsic value and for the health, social, economic, and cultural well-being of its peoples' (WWF, 1995). WWF in their comments on the drafting process, noted that the declaration establishing the Arctic Council called for making 'clear the integral relationship between environmental protection and sustainable development'. Otherwise, the Arctic Council could 'risk going in two directions, one cooperation for use of the Arctic, the other towards protection and sustainable use' (WWF, 1995).

In the fall 1995, the structure of the council consisted of two 'pillars' in the negotiations. One was the environmental work of the council, that is the AEPS, and the other the work on sustainable development. One of the pillars was called the 'Arctic Sustainable Development Initiative' (ASDI). Concerns were raised about what this kind of 'pillar thinking' meant for the future of the AEPS work (Elling, 1996). This was also seen in the name of a declaration. In Inuvik, Canada in 1996, it was time to draft a new declaration, this time it was titled the 'Inuvik Declaration on Environmental Protection and Sustainable Development in the Arctic' (1996).

The discourse in the negotiations made the protection and use of the Arctic environment and its resources a choice between two alternatives. First, the character of the Arctic environment was discussed. From the beginning of the AEPS, the Arctic environment was considered unique; it was understood that 'northern ecosystems are under greater stress than temperate regions' (AEPS 1991, p.17). The Arctic was considered different from other regions of the world. Most important, much of its population and culture is part of, and directly dependent on, the region's ecosystems, in particular its plants and wildlife. The uniqueness of the Arctic environment was threatened:

> Within the last few years we have become increasingly worried by the signs of degradation to the Arctic environment, which so far has been one of the cleanest and healthiest in the world (Olsen, 1991, p.1).

In early June 20, 1995 a draft on the commitment 'to protect and preserve the uniqueness of the Arctic environment' was reaffirmed (Arctic Council Declaration Draft, June 20, 1995).

The United States emphasised the recognition of Arctic uniqueness in order to protect

> ... the integrity of the aquatic, terrestrial, atmospheric and ice environments of the Arctic and their interdependent ecosystems as whole to the region itself and to the global environmental processes (Arctic Council Declaration Draft, October 20,1995).

This idea survived in the draft until the beginning of 1996. In the August 16, 1995 draft this uniqueness is directed towards both the indigenous peoples and the Arctic environment (Arctic Council Declaration Draft, August 16, 1995). As unique and worth of protection were the 'special relationship and unique contribution to the Arctic of the indigenous peoples' (Arctic Council Declaration Draft, January 15, 1996). The uniqueness of the indigenous peoples survived better than the environment in the end. There is no commitment to the preservation of Arctic environmental uniqueness in the final version. It is directed to the unique aspects of the Arctic and the special relationship and contribution of indigenous people and their communities (Declaration on the Establishment of the Arctic Council, 1996).

Second, the declaration balances between different interpretations of sustainability. In August 1995, the participants were ready to commit themselves to 'the sustainable use of its many resources in a manner that safeguards the Arctic environment and the well-being and cultures of indigenous peoples and other Arctic residents' and to the establishment of the new initiative known as the ASDI. The negotiations dealt with a broad definition of the emerging concept and this raised some concerns about the place of environmental issues and cooperation in the future collaboration. What is significant is the very broad understanding of sustainable development, including issues other than those that consider the role of environmental concerns in economic activities as a concept for sustainable development.

It was concern for the local and indigenous residents that held the attention: 'the Arctic requires special consideration in order that its indigenous peoples and the resident may obtain equal opportunities for social, cultural and economic development and emphasizing the responsibility of the Arctic states in this manner' (Arctic Council Declaration Draft, January 15, 1996). The two aims of cooperation were clear: 'safeguard the ecosystems and biodiversity of the Arctic region and facilitate sound management, including conservation and sustainable and equitable use of its natural resources consistent with the Rio Declaration, Agenda 21 and the Convention on Biological Diversity'; and to 'promote sustainable and equitable development in the field of economic relations and trade with a view to raise the standards of living, provide an adequate infrastructure, and ensure employment and the well-being of the indigenous peoples and other residents' (Arctic Council Declaration Draft, January 15, 1996).

After renegotiating this aspect in June 1996, the states affirmed their commitment to a 'sustainable development in the Arctic region, including economic and social development, improved health conditions and cultural well-being'. Then came the commitment to 'the protection of the Arctic environment, including the health of Arctic ecosystems, maintenance of biodiversity in the Arctic region and conservation and sustainable use of natural resources' (Arctic Council Declaration Draft, June 9, 1996). The states committed themselves to 'the well-being of the inhabitants of the Arctic' in the June 9, 1996 draft and establishing the human well-being as the main criteria for sustainability (Arctic Council Declaration Draft, June 9, 1996).

The final version affirmed the states' commitment to the well-being of the inhabitants of the Arctic. The second commitment the states made was to affirm their commitment to the 'sustainable development in the Arctic region, including economic and social development, improved health conditions and cultural well-being'. As a third objective, they affirmed 'concurrently' their commitment to, 'the protection of the Arctic environment, including the health of Arctic ecosystems, maintenance of biodiversity in the Arctic region and conservation and sustainable use of natural resources' (Declaration on the Establishment of the Arctic Council, 1996).

Third, in defining the relationship of the Council with the AEPS, the idea in June 1995 was for the Council to oversee and coordinate the AEPS (Arctic Council Declaration Draft, June 20, 1995). The important steps taken to develop and cooperatively implement the AEPS was recognized in the August 16 draft (Arctic Council Declaration Draft, August 16, 1995). The basis for a new working group or groups was made in the establishment of the Arctic Council. The August 1995 draft for the ASDI included 'working groups dealing with economic, social, cultural and other issues which may be identified' (Arctic Council Declaration Draft, August 16, 1995). In January 15, 1996-version, the commitment to the ASDI 'to deal with health, economic, social and cultural cooperation among Arctic states as well as with other issues for the well-being of Arctic peoples' was made (Arctic Council Declaration Draft, January 15, 1996). Commitment to the principles and goals of the AEPS was affirmed in the January 1996 version and 'the need to ensure that all relevant cooperative activities are in accordance with these principles and goals' (Arctic Council Declaration Draft, January 15, 1996). In the April 1996 version, the states recognized their commitment to 'the principles and goals of the Arctic Environmental Protection Strategy (AEPS) and recognizing the need to ensure that all relevant cooperative activities are in accordance with these principles and goals' (Arctic Council Declaration Draft, April 19, 1996). The latter part of the last sentence was dropped in the June 9, 1996 draft and it only reflected the recognition of the AEPS and its contributions (Arctic Council Declaration Draft, June 9, 1996). Only the contribution of the AEPS to the commitments of the Arctic Council was recognized in the final version. In this version, the task of the Arctic Council was to oversee and coordinate the programs established under the AEPS (AMAP, CAFF, PAME and EEPR) and adopt terms of reference for, oversee and coordinate a sustainable development program (Declaration on the Establishment of the Arctic Council, 1996).

Finally, through the idea of sustainable development states and indigenous peoples are making a claim on the future of the region. The normative order in the Arctic is now established around the concept of sustainable development. Sustainable development makes the use of the region and its natural resources possible - as opposed to defining the region as needing special protective measures. The priorities in the Arctic are sustainable economic development and pollution control. There are no signs that these priorities will change with the transition from AEPS-era to that of the Arctic Council. These tendencies are strengthened by the establishment of the Arctic Council.

The Arctic Idea of Sustainability

The Arctic Council negotiations have been processes of making the understanding of sustainable development richer in terms that included the broad issues of cultural, social, economic and health into sustainable development. The Arctic Council describes sustainable development as 'including economic and social development, improved health conditions and cultural well-being' (Declaration on the Establishment of the Arctic Council, 1996).

The language of governmentality is not only a question of meanings but also of different ways in making the world understandable and usable, where domains are made interventions for administrations, politicians, and experts. What is made doable in the Arctic is 'development'. The Arctic has emerged from the cooperation as a locale for cooperation rather than distrust. The militarized identity of the region was normalized through cooperation and the idea of sustainable development.

These definitions make the Arctic. They are constitutive meanings. Constitutive meanings - those assumptions, definitions and conceptions - structure the world in certain definite ways and constitute the logical possibility for the existence of a certain social practice. Constitutive meanings underlie social practices just as practices underlie actions. It is in terms of these meanings that the actors speak and act. Sustainable development is a modern understanding of the human-environment relationship. The Arctic Council establishes a modern mentality in the region. Modernity itself is a particular set of beliefs, a mentality (Foucault, 1984c, p.41).

Simultaneously, however, it was a process of making the concept of sustainability poorer. The Arctic Council was 'established as thin or deep as possible,' according to a close observer (Liljelund 1996). The mandate of Arctic Council considerably shrank during the negotiations. The first version of the tasks listed in the draft to establish the Council included fourteen items. According to the June 20, 1995 draft, the establishment of the Council would provide 'a forum to examine and discuss issues of common interest relating to the Arctic and to make recommendations pertaining to those issues'. The ideas for the work of the Council included 'to address the aspirations, concerns and objectives of the peoples living in the Arctic; to provide the political impetus for subsequent appropriate action by the Arctic governments on Arctic issues; to promote interaction among the Arctic governments and within the Arctic region in general to advance the programs of the Arctic Council: to advance Arctic interests by Arctic governments within appropriate international organizations' (Arctic Council Declaration Draft, June 20, 1995).

By August 1995, the long list was shortened to six objectives: '1) promote international cooperation and peace throughout the Arctic region, 2) provide a forum for addressing a wide range of Arctic issues, including the aspirations, goals and concerns of the Indigenous Peoples and other Arctic residents, 3) safeguard the uniqueness of the Arctic environment and promote the sustainable use of the natural resources in the region bearing in mind the principles and recommendations set out in the Rio declaration and Agenda 21, 4) advance the principle of sustainable development including its economic, environmental, social and cultural dimensions, 5) provide political impetus for appropriate action on Arctic issues by the governments of the eight Arctic countries, and 6) enhance the collective environmental security of Arctic states and peoples' (Arctic Council Draft, August 16, 1995).

The final version divided the work of the Council into only four tasks: to provide a means for promoting coordination and interaction among the Arctic states, with the involvement of the Arctic indigenous communities and other Arctic inhabitants on common Arctic issues, in particular issues of sustainable development and environmental protection in the Arctic; oversee and coordinate the programs under the AEPS; adopt terms of reference for, and oversee and coordinate a sustainable development program, and disseminate information, encourage education and promote interest in Arctic-related issues (Declaration on the establishment of the Arctic Council, 1996).

Economic Ethos

Through the idea of sustainable development indigenous peoples are redefining their identities as economic actors. Sustainable development in the Arctic means that the indigenous peoples are increasingly seen as economic actors. The economic concerns of indigenous peoples became evident in the work of the TFSDU. For example, in the discussion on trade barriers to Inuit products, the problems were defined as legislative barriers such as Marine Mammal Protection Act of the United States (MMPA) and the European Union (EU), psycho-sociological and cultural life study barriers by the consumer, socio-economic barriers, international barriers and people's attitudes. The most notable trade barrier, according to a Danish discussion paper, is the MMPA of the USA. The Act dates from 1972 and prohibits the import of all and any marine mammal products into the USA. The MMPA prohibits the import of all whale products from Arctic countries (Trade Barriers Affecting Products from the Arctic, 1995; see also ICC, 1995b).

The market for marine mammal products in Europe has been partially closed by the 1983 Seal Skin Directive. This directive prohibits the import of products derived from the pups of harp and hooded seals. The directive makes an exception for products deriving from the Inuit subsistence seal hunting. Individual EU countries have, however, developed their national seal skin legislation. The import of fur products into the EU from different Arctic and sub-Arctic species will be prohibited, unless the countries exporting these products forbid the use of leghold traps, according to the EU plans. The discussion paper calls on the 'world community to respect the legitimate interests of the Arctic peoples, and take the necessary measures to protect the international trade in sustainably harvested products of nature' (Trade Barriers Affecting Products from the Arctic, 1995).

In promoting sustainable development in the Arctic the cooperation has resulted in common action aimed at improving the situation of the hunters in indigenous peoples communities. Canada, for example, made a draft letter of intervention to the EU about the seal skin products ban. Canada sent a verbal note to the EU. Denmark said that it would support Canadian approach to the EU at the task force meeting in March 1995, but support was also required from other countries. It was agreed that the U.S. and Russian Governments could draft similar letters of intervention using the Canadian letter as a model. Task Force countries that are members of the EU would apply internal pressure (TFSDU, 1995a, p.10).

Economic issues, not only those related to the economic well-being of the indigenous peoples, in the Arctic may also be of common concern for the states in the future. The structure of the Arctic Council suggests the possibility of such concerns. The mandate of the Arctic Council definitely brings economic issues and the welfare of the Arctic populations into discussion. As Foucault suggests, the ultimate goal of governmentality is 'with how to introduce economy, that is the correct manner of managing individuals, goods, and wealth ... into the management of the state' (Foucault, 1984a, p.15).

These discourses are all carried out in the name of sustainability. Governmentality includes the ways in which individuals may redefine their existences. The idea of self-subjection opens the possibility of identity politics; identities are fashioned by the political technologies of individuals that totalize as they individualize. Foucault is interested in the process of identity formation; becoming oneself and the practices of self. Therefore, it is essential to examine those intentional and voluntary actions by which men not only set themselves rules of conduct but also seek to transform themselves, to change themselves (Foucault, 1982b, p. 237-238).

Identities are made by making ethical claims in relation to the concern about the environment. It is a question of 'the kind of relationship you ought to have with yourself ... and which determines how the individual is supposed to constitute himself as a moral subject of his own actions' (Foucault, 1982b, p.238). This is called 'ethos': it is a way of a social unit speaking to itself about itself and constituting itself as a result. Ethos does not separate fact from value nor the subject from the object of knowledge. An ethos means a feeling, an identity and a commitment. It is about values, the root for the word being 'ethical'. It is about both knowledge and feeling. An ethos links the claims of truth and value. It is a social self, a shared self and a role. It does not infer anything artificial; it is rooted in the subject at issue, and expresses the commitment of the speaker (Myerson and Rydin, 1996, p.23).

Ethics and subjectivity are inseparable; to be a subject is to have ethos.It is an ethical practice that defines moral obligations. Ethos and subjectivities are formed in discursive moralities. These do not have absolute foundations nor they are based on compromises. They are concrete, local and contingent. Ethics is historically grounded social practice (Hekman, 1995, p.149). Ethos is a product of collective self-interpretations and self-identification of human communities. They are different from the notion of a consensus or a compromise about values or beliefs. Ethos is intersubjective: in the Foucaultian sense intersubjective meanings themselves are products of the long-range subjectifying trends in our culture (Dreyfus and Rabinow, 1982, p.165).

5 Governmentality in the Arctic

The Meaning of Environmental Cooperation

The Practice of Cooperation

It can be argued that the status of cooperation does not really correspond to the assumptions of cooperative effectiveness on environmental issues expected by the standard regime approach. Disappointment, here, is obvious and unavoidable. The AEPS is programmatic rather than regulative in character; The impact of the AEPS on environmental problem solving has been 'modest'. Not everyone, however, shares this view. According to Håken Nilson, 'cooperation on concrete environmental issues has been strengthened' by the AEPS experience. The main success of the AEPS is that 'it has provided a mechanism for the Arctic states to initiate and maintain a constructive dialog, and for a significant increase in the knowledge about the Arctic environment' (Nilson, 1997, p.4).

It is, however, interesting that so much effort has resulted in so little, which leads to the question of what is the point of the effort. Oran Young notes that 'it would be a mistake to seize on these limitations to dismiss the AEPS out of hand as an ineffective and largely irrelevant effort at environmental cooperation'. So far, cooperation has produced 'a vibrant social practice that is engaging the interest of a wide range of public agencies as well as non-state actors interested in Arctic environmental concerns' (Young, 1995a, p.8).

Cooperation in the Arctic has produced a practice of a series of ministerial meetings and working groups getting together. The AEPS participants seem to think that developments in cooperation have been successful. The Finnish view was that the cooperation has been 'extremely successful' in terms of awareness of the problems of the region (Haavisto, 1997a, p.3). The Canadians, so far, view the AEPS as a dynamic cooperation It is in a stage of transition from 'analysis and monitoring, toward a stage where concrete actions will be come more prominent in our work'. In addition, environmental cooperation is integrated into a wider framework which, in the future, will gradually encompass many other activities. A greater understanding and awareness of the challenges that face the region have been gained through the

AEPS. In the future 'the circumpolar states will need to work hard within the broader global community to achieve effective results' in dealing with pollution issues (Simon, 1997, p.2).

The Swedish saw the AEPS as, an 'example of how international cooperation founded on common interests, goals and commitments, rather than legal instruments, can be successful and constructive' (Kjellen, 1997). The Icelandic representative 'stressed the need to maintain momentum to protect the Arctic environment'. It was hoped that 'the next state for Arctic cooperation will be more characterized by concrete actions and less by analysis and monitoring' (Bjarnason, 1997). The Russians seemed ready to include as many countries as possible in Arctic affairs and not 'just the parties to the AEPS/Arctic Council' (Solovyanov, 1997).

According to the Norwegian view, the 'AEPS has helped strengthen the political dialogue among the Arctic states and the setting up of the Arctic Council is a significant milestone'. The Norwegians emphasized that 'environmental protection remains an essential and integral part of the Arctic Council's future activities and priorities' (Berntsen, 1997). Also the statement by Tucker Scully called for clear priorities in cooperation (Scully, 1997). The WWF notes that the Alta ministerial meeting 'is not only a historic step forward in the process of environmental cooperation'.

> Six years have passed since the Arctic countries met in Rovaniemi at the initiative of Finland. What they created - an international cooperative process which changed the Arctic from a zone of confrontation to one of environmental cooperation - is a tremendous achievement in the world (WWF, 1997, p.1).

The WWF compliments the Arctic environmental cooperation for moving toward, 'a more action-oriented agenda' (WWF, 1997, p.2).

Negotiating at international level itself is a social practice. As Friedrich Kratochwil suggests, 'actors are not only programmed by rules and norms, but they reproduce and change by their practice the normative structures by which they are able to act, share meanings, communicate intentions, criticize claims, and justify choices'. Therefore, Kratochwil continues: 'one of the most important sources of change, neglected in the present regime literature, is the practice of the actors themselves' (Kratochwil, 1989, p.61).

These practices more resemble habits than legal rules. As Rosenau (1986, p.861) claims actions stem, 'from a combination of past experiences, cultural norms, memories, beliefs, personality, role expectations and cognitive styles to which they have long been accustomed, by which they manage to maintain continuity in their affairs'. In Rosenau's view, it is a world of habit driven individuals, collectivities and their organizations. The world is of 'our making' through application of practice-based rules (Onuf, 1989, p.73). In an analysis of the Arctic environmental cooperation, Carola Björklund suggests that even if cooperation had taken a less codified form, in practice the legal form is not decisive for the legitimate protection of the environment in the polar areas (Björklund, 1995, p.146).

NGOs in the Cooperation

In particular, the field of international environmental politics seems to attract non-governmental organisations (NGO) or non-state actors. It is often assumed that the growing role of groups in international relations is the result of the effect interdependence has on intrasocietal, national and global levels. The world is increasingly interconnected, thus providing 'transnational actors' with the potential for unprecedented influence if they mobilize effectively. This is not to suggest that transnational actors operate without constraints but rather to emphasize the 'fluidity' of global political relations under conditions of interdependence (Willetts, 1982; McCormick, 1989; Porter and Brown, 1991; Weiss and Gordenker, 1996).

The effort to give a name to the phenomenon of the emerging role of actors other than states reflects 'a more complex international society in which states remain important actors, but find themselves increasingly sharing influence, if not authority, with several other types of actors' (Young, 1990, p. 344; see also Keohane and Nye, 1977, p.24-25). Thomas Princen points out that NGO influence is exerted by linking the local to the international levels of politics - that is interpreting interdependence. It is influence gained by filling 'a niche that other international actors are ill-equipped to do' (Princen, 1994, p.41).

The 'non-governmental' and 'non-state' is often evaluated as problematic: how can something be defined by using the word 'non'? James Rosenau suggests the term NGO unduly places too much attention on states and thus maintains state-centrism in international relations scholarship. He suggests using the term 'sovereignty-free actors' making a division between non-state actors and states which Rosenau calls 'sovereignty-bound' (Rosenau, 1990, p.36). However, changing the name as suggested does not deal with the problem. It still separates NGO's, or sovereignty-free actors, from international politics and avoids asking the question about their role. Sovereignty is an inherently relational notion making the states the main concern for the study at the expense of other collectivities and actors that may be significant internationally, such as 'non-governmental' or 'non-state' actors. Questioning this definition means challenging the neoliberal view on interdependence and domestic politics:

... interdependence theorists could not have it both ways; either they were right in their talk about blurring, with the inevitable consequence that their theories ceased to be theories of international politics, or as was more often the case, talk about blurring was mere lip-service (Bartelson, 1995, p.20).

This means 'that which is blurred essentially is distinct; in the end one was tacitly reaffirming the same distinction which one so valiantly criticized' (Bartelson, 1995, p.20). Sovereignty as a theory and practice establishes the boundary between what is internal and what is external; it is a practice of constituting political reality for the agents themselves whose identity in turn depend on this division (Bartelson, 1995, p.41).The name itself - 'non-governmental' or 'non-state' actor - creates a boundary between states and others and reaffirms the assumed boundaries.

The study of governmentality results in interest in the identities and roles defined by regimes including power considerations. Foucault's considerations on power opens discussion on the participants of cooperation and regimes in an interdependent situation. According Foucaultian understanding, power is always exercised in relation to rules which impose the agent as a social actor. The individual assuming a role is an effect of power and, while to the extent to which it is in effect, the element of its articulation. For Foucault, acquiring roles and positions is an issue of institutionalizing the sites of discourse. Foucault speaks of discursive practice. Discursive practice is a body of anonymous, historical rules, always determined in a time and space that have defined a given period and for a given social, economic, geographical or linguistic area and the conditions of operation of the enunciative functions. This enunciative field does not refer to an individual subject nor to some kind of collective consciousness or transcendental subjectivity. It is described as an anonymous field whose configuration defined the possible positions of speaking subjects (Foucault, 1972, p.118-119).

As Foucault concentrates on the relationship between states and individuals and the processes by which individuals receive and construct identities, the social or collective dimension of identities is mostly put aside. The individuals involved in such conflicts may become more preoccupied with asserting their own identities than with other political goals, such as establishing solidarities with other groups. Based on such identities we recognize ourselves as members of a social group or state. The struggle for one's own identity links with other more collective struggles against the government of individualization (Foucault, 1982a, p.211-212).

However, there is freedom in the constitution of subjectivities with a capacity for critical independence or self-governance (McNay, 1992, p.104). The capacity of movements, such as the activities of indigenous groups, depends on their ability to construct a powerful identity and effective strategies and the given opportunities in society to defend or advance the ideology of the movement (see Hjelmar, 1996, p.177-178; Touraine, 1981, p.29-30; see also Wendt, 1996, p.51).

The Inuits wanted to become involved in the AEPS at an early stage. Mary Simon who then represented the ICC at the Yellowknife meeting said that the 'ICC made a decision to expand cooperation to this level' and participated in the Yellowknife meeting (Simon, 1996). Simon noted that a growing number of issues affecting Inuit rights and interests are increasingly being regulated at the international level. 'National and regional initiatives alone are not adequate to protect Inuit communities'. Simon concluded that 'there must be a significantly expanded role for Inuit at the international level' at the conference organized to formulate the Arctic policy in 1985 (Simon, 1987, p.37-38; see also Clark and Dryzek, 1987, p.227-228).

The three organizations, especially the ICC, have worked to develop an identity as an expert on Arctic issues. Rosemarie Kuptana, president of the ICC noted that 'language surrounding our involvement in the AEPS is often couched in phrases such as the governments have finally involved the indigenous peoples in the important environmental work'. She stresses an opposite view:

> Our participation is based on our interest. It is simply natural that we are involved. We have a strong commitment to this work because it is carried out in our homelands and because it is within us (Kuptana, 1996, p.1).

The ICC has successfully followed this advice in the AEPS and the Arctic Council. The ICC used a strategy composed of: 1) a political program stating the aims of the organization, 2) constructing a clear environmental identity, 3) having both intellectual and material resources to contribute to the process by preparing reports and statements on different AEPS activities, and 4) making and maintaining contacts with the Canadian government and other AEPS participants. The ICC developed an Arctic policy in the mid-1980's. This is a comprehensive Arctic policy in Inuit circumpolar regions with regard to matters of economic, social, cultural and political concern (Principles and Elements for a Comprehensive Arctic Policy, 1992, p.31).

Next, the conservationist identity of the group was advanced by the development of the Inuit Regional Conservation Strategy. The environmental commission of the ICC was established in 1985. The commission developed the framework document Inuit Regional Conservation Strategy (IRCS). This is a common environmental strategy for sustainable development covering Alaska, Arctic Canada and Greenland (Towards an Inuit Regional Conservation Strategy, 1986, p.5). The Inuit Regional Conservation Strategy received a Global 500 award from the Secretary General of the United Nations in 1989 (Lynge A., 1992, p.7).

Then, within the AEPS, the ICC has been productive in offering their reports for use by the working groups. The reports include two volumes on traditional knowledge submitted before the Nuuk meeting and Regional Agenda 21 from the Inuit Perspective. It also has two research projects within CAFF and TFSDU: collecting traditional knowledge on beluga whales in Alaska and a retrospective study on the seal market collapse in North America after the EU ban (see TFSDU, 1995a). The role of the ICC could not have been possible without the support of the Canadian government. At the SAAO meeting in Iqaluit in 1995, Canada stated that 'it wanted to achieve a partnership in Arctic environmental management which included Indigenous Peoples' Organizations at a management rather than observer level' (SAAO, 1995, p.9).

The former ICC president Mary Simon was appointed an Arctic Ambassador of Canada in 1995. As the Arctic Council is a Canadian initiative, her role has been central in the negotiations. Rosemarie Kuptana was the co-chair for the Arctic Council Panel which wrote the NGO proposal for the Council and later, as the President of the ICC, participated in the negotiations to establish the Arctic Council. Chester Reimer, former research director of the ICC, worked for a while as the executive director of the IPS in Copenhagen. Henry Huntington explains the success of the ICC: 'it has a good representation, well respected' and it is making 'substantive contribution' (Huntington, 1996). Mary Simon explains the leadership role of the ICC by its intellectual and material resources. She also refers to the input of individuals. The basis of the 'leadership role of the ICC' is funding (Simon, 1996). The ICC is described as better organized compared with other indigenous groups (Fenge, 1995). The distribution of work between national Inuit organization and the ICC, which concentrates on international concerns, is important for the success of the ICC, too.

Finally, the success is based on having connections at different levels. Jorgen Sondergård, for example, explains that 'Greenland Home Rule takes part in HOD [Head of Delegation] meeting where the indigenous peoples' organizations cannot be'. The cooperation is organized informally: 'We don't hold formalized meetings [but] on spot ad hoc meetings when needed'. The good contact between the ICC and the Greenlandic Home Rule Government helps 'the flow of information' (Sondergård, 1996).

Sámi efforts at constructing an expert identity are not easy to find - not even within the AEPS. Ritva Torikka explained that her activities as the representative of Sámi in some AEPS activities had been difficult since there was a 'lack of clear political goals set by the Sámi Council'. The environmental program of the Sámi Council accepted in 1986 was not enough (Torikka, 1996).

The Sámi representatives also explain that 'the Sámi do not have enough resources for a meaningful participation'. However, even when travel funding has been available there has been no Sámi representative participating in meetings. The ICC has personnel for the advance preparation for meetings, but such possibilities are not open to Sámi representatives. Sámi representatives to negotiations participate in the meetings on a volunteer basis, not ex officio. The Sámi did present their projects to advance indigenous knowledge, but not as successfully as the ICC. The representative of the Sámi Council said that 'we have to define our projects,' but he did not doubt that the Sámi could contribute: 'We have people who be can there'. According to Halonen the problem is funding for developing projects in collecting and using Sámi knowledge (Halonen, 1996a). The lack of funding has limited Sámi participation in different AEPS meetings: 'It is not possible to participate even in all the meetings for the lack of resources in the Sámi Council,' according to Torikka (1996). However, it seems that the Sámi Council is content with the existing situation; Leif Halonen from the Sámi Council explained the situation by stating 'having involvement and to be heard is a goal itself' (Halonen, 1996a).

The situation in the Association of Indigenous Minorities of the North, Siberia and Far East of the Russian Federation has been complicated. Many of the representatives of the Russian association have been satisfied with only observing the negotiations with the aid of an interpreter. However, not much material has been translated into Russian and this has made active participation difficult. Finally, the unclear identity of the association has made things difficult. There is a disagreement on whether it is a social or a political organization. Lately, the organization has suffered from internal problems. At the beginning of the organization there were disagreements between indigenous peoples' representatives over the need for the organization; the founder, Sanghi, explained these as being caused by 'puppets of the Soviet system' (Alia, 1991, p.29). The internal disagreement seems to reflect Sanghi's disappointment at not being reelected as president of the organization. The former president has established a new organization to represent the interests of the northern peoples. This has resulted in confusion over representation for the Russian indigenous peoples in the AEPS (personal communication an.).

Excluding Environmental NGOs?

The Swedish representative noted in 1991 that the cooperation:

... is an open process. We must continue to provide a participatory role, as observers, for non-Arctic countries and international bodies which are willing and able to make significant contributions to our efforts (Dahl 1991, p.8).

She continued: 'environmental NGO's, with particular interest of expertise in Arctic matters, must be also given opportunities to contribute and participate in our work' (Dahl, 1991, p.8). Environmental NGO's have, however, been rather invisible in the cooperation, compared with many other international negotiation processes and the role the IPO's have acquired in the cooperation. According to Dave Cline (the NGO delegate of the U.S. delegation at the 1991 meeting) from the National Audubon Society, environmental organizations have often shown 'little interest in the Arctic' (Cline, 1996; see also Hurwich, 1995; McCloskey, 1995). There are two exceptions: the IUCN and the WWF.

The International Union for Conservation of Nature and Natural Resources (IUCN) has had an interest in the Arctic for a long time. The IUCN had already been suggested at the World Conservation Strategy (1980) that the Arctic environment and its problems should be taken into consideration.

The World Wide Fund for Nature (WWF) established an Arctic coordination bureau in Oslo, Norway in 1992. The WWF, with its national organizations in most of the Arctic countries and projects in Russia, has been interested in building a pan-Arctic non-governmental lobby for Arctic conservation. The tasks of the WWF Arctic program, defined by the Arctic coordinator Peter Prokosch, were to 'create a common pan-Arctic thinking,' as well as acting on and supporting conservation steps and developments in the Arctic. To achieve that goal, the WWF aims to develop assessments of natural values, classified threats and conservation projects. Solution models for environmental problems have to be visualized and mapped. The main task of the WWF is, however, to form a circumpolar lobby for the Arctic nature as a whole (Prokosch, 1992, p.6).

The WWF described itself as 'the only nongovernmental organization with a truly circumpolar scope' (Prokosch, 1997). An important contribution of the WWF was the Arctic Bulletin, which for a long time was the only source of information about the AEPS activities. The WWF participated as an ad hoc observer at the 1996 Inuvik meeting. It has also participated in the activities of the CAFF working group. In its application for observer status within the Arctic Council, the WWF argued that it had demonstrated 'a deep commitment to the health of the Arctic environment and the Arctic residents who depend upon this environment for their livelihood' (Prokosch, 1997). In the current situation these kinds of organizations such as WWF do not seem to have a place within the framework of the Arctic Council. The participation of the WWF in Arctic Council activities has lately been questioned for defining the procedural rules of the Arctic Council activities (Arctic Council SAO, 1997, p.29).

Most environmental NGO participation has taken place through national delegations. There has been an NGO representative in the U.S., Canadian and Norwegian delegations. Both the National Audubon Society (David Cline from the Alaska office) and the U.S. Arctic Network (Margie Gibson) have been represented in the U.S. delegation. The Canadian Arctic Resources Committee (CARC) has participated in the work of AEPS for the Canadians. CARC is a public interest and charitable organization established in 1971. Its task is to expand the concept of national interest in the north to include the rights and interests of indigenous peoples, to transfer authority to the local level and the conservation of natural places. CARC was appointed as a member of the Canadian Delegation to the AEPS ministerial meeting in Nuuk in 1993.

CAFF has been the forum most welcoming to environmental NGO's (see AEPS, 1995). Such organizations as Birdlife International and the World Conservation Monitoring Center have been present at its meetings. Birdlife International was established in 1993 and it is built on the work of the International Council for Bird Preservation, which was established in 1922. This is a federation of conservation organizations and describes itself as 'non-confrontational, working together with governments' (Heath, 1994, p.95-96). The World Conservation Monitoring Center is a joint venture between IUCN, the WWF and the United Nations Environment Program (UNEP). It has been involved in the compilation of global environmental data for the CAFF program (Kaitala, 1994, p.91-92). The UNEP contribution has been through its GRID center in Arendal, Norway. The center has produced the maps for AMAP use (see Kullerud, 1994, p.92-93).

The International Conservation Union has also participated in CAFF meetings. It is a non-governmental organization 'dedicated to the ecological and cultural integrity of the Arctic for present and future generations'. The aim of the organization is to link 'environmental interests, indigenous peoples' interests and sustainable development interests with women's and other networks around the world who are interested in building an international Arctic constituency' (Hurwich, 1994, p.103).

It appears that the strategy suggested by the International Conservation Union is the only one that provides a way for Arctic environmental cooperation. The problem is that the most environmental NGO's lack an 'Arctic constituency' and long-lasting interest in northern transboundary environmental issues (Cline, 1996). Many environmental NGO's are at least perceived as representing the concerns of environmental activists in the southern centers. These organizations have shown very little sympathy to, and understanding of, the concerns of indigenous peoples (see Lynge F., 1992; Wenzel, 1991).

An interesting organization is the U.S. Arctic Network which represents 25 organizations, including native and conservationist groups. The objectives of the association are to: promote conservation of the circumpolar Arctic ecosystem; protect indigenous cultures and ensure the sustainability of local communities; foster understanding and cooperation between individuals and organizations that share common goals for the protection of the Arctic; to empower indigenous and local peoples, whose lives are an integral part of the Arctic ecosystem, so that their participation in establishing an Arctic policy is meaningful and effective. The steering committee of the Arctic Network includes representatives from: the Native Migratory Bird Working Group, Greenpeace, the Environmental Defense Fund, the Rural Alaska Community Action Program, the Eskimo Walrus Commission, the Aleutian/Pribilof Association, The Wilderness Society and Tanana Chiefs Conference. The network changed its name to the 'Arctic Network' after establishing an office in Providenya, Kamchatka (Chukotka) (Gibson, 1996).

The environmental NGO's have been less successful in their effort to create a place for themselves in the Arctic cooperation. Steve Breyman defines the relationships between social movements and knowledge: 'Knowledge can be used in the pursuit of movement aims to... monopolize or share claims to meaning' (Breyman, 1993, p.128). The power of the indigenous peoples and in particular their organizations are in 'their persuasive capacity' (Sondergård, 1996). Indigenous knowledge is inseparable from the people themselves; it requires the direct participation of indigenous experts.

The establishment of the Council reaffirmed the positions that the different actors had gained in the discourse. In the rather long and difficult discussion on the categories of participation and representation the states have aimed to create practices for participation. These practices contain ideas on who has the right to speech, from which institutional places the speech can take place and what positions the objects of the discourse can have. Through creating different categories of participation and limiting the participation by criteria, such as 'substantial contribution,' the access of different interested actors in the Arctic is restricted. The practices of states maintain the boundary between state actors and non-state actors.

The Meaning of Establishing the Arctic Council

The Arctic Council as an Organization

The current plans for the Sustainable Development Programme include 'cooperative activities' meaning 'a particular activity of any type authorized by the Arctic Council to be carried out under a Programme of Work, including activities of Working Groups, Task Forces, or other bodies established by the Arctic Council'. The Arctic states may establish working groups, task forces, or other bodies to carry out programs and activities under the guidance and direction of SAO's. The rules focus on the administration of such a body; it may select a chair and a vice chair or a lead country may volunteer to provide chair and secretariat support. The period of which a chair or a vice-chair may sit shall be specified. The procedures give advice on the date, location and agenda for the meetings of working groups, task forces, and other bodies. They shall be decided by 'a consensus of the participating Arctic states, after consultation with the representatives of the participating permanent participants' (see Arctic Council SAO, 1997).

The rules for procedure say very little about the actual content of the proposals that can be made. The cooperative efforts are to be initiated by an Arctic state or a permanent participant by submitting a proposal. A proposal shall include:

> ... the issues or matters to be addressed, the reasons, for the Arctic states to consider for approving the proposal, any recommendations of SAOs in relation to proposal, including recommendations as to an appropriate body or bodies for carrying out, coordinating, or facilitating an activity and any special rules or procedures that should apply to such body or bodies (Arctic Council SAO, 1997).

A proposal has to include information on the costs and methods of financing an activity, a working plan with dates for completion, information on relationships with other Arctic Council programs and activities and to other relevant regional or international fora, and any other information relevant to the proposal (see Arctic Council SAO, 1997).

The future work of the Arctic Council will define the content of sustainable development. When the working group was established, the Finnish representative noted:

> Maybe we have concentrated too much on the formal status of sustainable development in the AEPS and in the forthcoming Arctic Council. Now we

have reached the consensus that the working group status is the most functional one. Afterwards much more attention should be paid to the process of giving substance to Arctic sustainable development (Nurmi, 1996, p.3).

Giving substance to sustainable development will not be an easy project; According to the Finnish proposal, 'the Arctic Council would be a suitable forum to coordinate the Arctic implementation of both Agenda 21 and the priority areas set by the UN Special Session of General Assembly' (Nurmi, 1996, p.3). The Finnish wish is for an Arctic Agenda 21 to be prepared. This is nothing new in itself. The idea of creating an Arctic Agenda 21 under the the prgoramme of TFSDU has essentially been forgotten in the struggles to decide the structure and content of the work of the Arctic Council (Rouhinen, 1996; Matero, 1997).

Understood as a bureaucratic structure, the Arctic Council hardly exists as an organization.the 'new institutionalist' approach focuses on specific institutions; it emphasizes international regimes and formal international organizations such as formal intergovernmental or cross-national nongovernmental organizations, international regimes, and conventions (Keohane, 1989, p.3-4). The idea of establishing the Arctic Council was to strengthen existing cooperation. One of the most important substantive issues to be dealt in the Arctic Council was, according to the Arctic Council Panel (1991), the protection of the Arctic environment. The Arctic Council Panel stated that the weakly institutionalized Rovaniemi process would be better subsumed into the agenda of an Arctic council. This may, in turn, serve to energize the Rovaniemi process on specific issues (Arctic Council Panel, 1991, p.2-3).

The establishment of the Arctic Council did not bring new bureaucratic structures into existence. It did not bring about a permanent secretariat. This would have really made a practical difference for secretarial support and provided possibilities to advance the cooperation and coordination of the different working groups (Skåre, 1996). The states did not appear dedicated to Arctic environmental commitment for the development of the organization of the work.

Cooperation had suffered from a lack of funding and that lack continues. A chronic shortage of funding has been a reality for the AEPS. In a preparatory meeting in Kiruna in 1991, it was noted that 'the preparations for future meetings would be the responsibility of the host government' (Protecting the Arctic Environment, 1991, p.4). Funding for the AEPS secretariat which coordinates AEPS meetings and produces minutes of meetings operates under the host country principle. A host country pays the entire cost of the AEPS secretariat during its tenure. Four countries, Finland, Greenland, Canada and Norway have acted as host countries. Two of the working groups, the AMAP and CAFF, work under the host country principle. Their rotation among host countries is not regular (Smith, 1997, p.8).

For the time being the plan is to maintain principle of voluntarism. Secretarial services are, however, covered by obligatory fees. In particular the Russian participation was discussed. All possibilities of cost-sharing and of Russian contributions in kind was considered worth examining (see The Arctic Environment, 1993, p.15). The ministers in Nuuk agreed to 'seeking resources to enable each country to fully participate in the program activities under the Arctic Environmental Protection Strategy' (The Nuuk Declaration on Environment and Development, 1993).

The WWF claims that the AEPS practice has been: 'each country has decided how much it will pay and even whether it will pay at all' (Smith, 1997, p.8). No progress was made in Inuvik in 1996; the states agreed to 'ensure implementation of the priorities listed in the present declaration and to make every effort to provide and maintain the necessary resources to enable each country and indigenous peoples to participate fully in the activities of the AEPS' (Inuvik Declaration on Environmental Protection and Sustainable Development in the Arctic, 1996). The states establishing the Arctic Council agreed to 'regularly review the priorities and financing of its programs and associated structures'. It was also agreed that the responsibility for hosting the meetings of the Arctic Council, including a provision for secretariat support functions, should rotate sequentially among the Arctic states (Declaration on the Establishment of the Arctic Council, 1996). At the Alta meeting in 1997, the issue of financing activities was postponed.

For indigenous peoples, Arctic cooperation has meant a chance to strengthen the organizational setting of their collaboration. The Indigenous Peoples Secretariat (IPS) was established in 1993 on the initiative of Denmark and the Greenland Home Rule Government to 'address all issues related to the participation of indigenous peoples'. The aim was to enhance the participation and contribution of indigenous peoples in the conservation and protection of the Arctic environment, and to bring their knowledge to bear on these matters (IPS Information, 1996).

The role of the IPS is to facilitate a dialogue between the three organizations and the governmental bodies of the AEPS. The Governing Board of IPS comprises six members. Three of them represent the three IPO's. The three remaining members of the Governing Board are government representatives: Denmark (permanent member), the presiding host of AEPS country and a representative from the one of the remaining six Arctic states. The eight Arctic states appoint this member. One of the three IPO's chairs the Board. The Governing Board meets at least annually (Report of the Third Ministerial

Meeting, 1996, p.19).

At the beginning, organizational problems hampered the work of IPS. However, nobody denies its useful role in coordinating the work of the IPOS; the 'indigenous people's secretariat is a helpful addition' (Huntington, 1995). The problem with the IPS is that its status is unclear; it does not have an international and legally binding foundation. This would ease some operational difficulties that the secretariat has (Petersen, 1996). However, 'it has potential', says Chester Reimer, the former executive director of the IPS (Reimer, 1995).

The wish of the Arctic indigenous peoples came true with the establishment of the IPS. Representatives of the Arctic peoples of Canada and Scandinavia met in Copenhagen in 1973. The conference discussed the idea of establishing a permanent circumpolar international organization. It proposed forming a 'Circumpolar Body of Indigenous Peoples to pursue and advance our shared and collective interests'. The working committee established at the Arctic Peoples' Conference did not succeed in organizing a new conference, partly due to lack of financial support and partly because the persons involved were engaged in other time-consuming activities. The participants in the conference represented many indigenous groups in the Arctic (Kleivan, 1992, p.233).

The practice of financing the participation of indigenous peoples' organizations reveals another aspect of the story. According to a report by the IPS, during the first three years (1991-1993) the involvement of the Arctic indigenous peoples was sporadic and funding was 'extremely limited'. Financing IPO participation in the first period was based on ad hoc arrangements. In the latter period (1993-1996) certain countries (Canada, Denmark, the United States and Norway) provided funds to indigenous peoples living in their own countries. Iceland and other governments provided some funding and services-in-kind for the AEPS indigenous knowledge seminar held in Reykjavik 1995. Denmark and Canada also created funds to assist indigenous peoples in Russia (Reimer, 1996, p.7). The special responsibility of the federal government does not obligate the U.S. State Department to handle the practicalities of participation, such as financial support for the indigenous groups (Hild, 1996).

The IPS report stated that for the Canadian Inuit most funding had been received from the Government of Canada. The total amount was 175,000 dollars per annum. Other project funding related to AEPS activities had also been provided. The ICC Head Office had, from 1995 till March 1997, received 25,000 dollars for preliminary Arctic Council work. For the Greenlandic Inuit most funding was given by the Danish Government. Since 1993, the average annual funding has been 70,000 dollars. The Alaskan Inuit have received 25,000 dollars per annum since 1994 (Reimer, 1996, p.5). Russian Inuit have attended some meetings of the Arctic Council. These costs were covered by the ICC Canada and the Government of Canada. Both Canada and Denmark established funds to assist the Russian indigenous peoples' organization to have representatives at some AEPS meetings. Canadian government financing for Russian indigenous participation, administered by the ICC Canada, is 30,000 dollars per annum. The Danish Fund provides travel expenses for AEPS activities. These funds have been 'underutilized' (Reimer, 1996, p.6-7).

Only Norway has provided funding to the Sámi Council for participation in the AEPS. Norway has granted 4,000 dollars per year. Finland, Sweden and Russia have not contributed directly to the Sámi Council for AEPS work. Some funding for an AMAP project has been forwarded to the Sámi Council. This lack of financial support has resulted in AEPS meetings not being officially attended, being attended by volunteers when available and in a lack of overall coordination of Sámi activities in the AEPS. It has also resulted in a major deficit for the Sámi Council which has contributed from its core budget to AEPS activities (Reimer, 1996, p.6).

In this respect the existing Arctic Council is far from the original concept presented by the Arctic Council Panel. This suggested that 'adequate financial support will be provided by the Arctic states to permit aboriginal organizations and communities to take part in the word of Council and Working Groups' (Arctic Council Panel, 1991, p.28).

Northern Cooperation on Sustainable Development

Sustainable Development is strengthened by other cooperative arrangements in the region. The Northern Forum has served as a mechanism for regular international interaction between northern leaders since 1990. The purpose of the forum is to 'improve the quality of local, national and international decision-making regarding northern issues by providing a means through which northern voices can be heard at all stages of the process'. The board of directors consists of governors and high political officials representing the northern regions. Associate membership is available to businesses, universities and special interest groups (Statement of Intent, 1990, p.2). According to the Northern Forum Secretariat, indigenous peoples in the north are represented through governors in the Northern Forum (Clark, 1996).

The Northern Forum also has environmental projects such as on environmental monitoring and wildlife management (Clark, 1996;Wohl, 1996). The Northern Forum, an organ describing itself as 'Northern voice' also applied for special status within the Arctic Council structure. The basis of special status for the Northern Forum is that 'unlike other organizations with an interest in observing the work of the Arctic Council, the Northern Forum is the only organization which deals with all these issues and works with all the constituences of the northern regions' (Eriksen, 1997).

The main concern of the Northern Forum is the economic development of the people in the north. The Tromsø declaration emphasizes that 'we have the right to use renewable resources responsibly for subsistence purposes, to develop an economy and to continue our unique lifestyle' (Tromsø Declaration, 1993). Sustainable development in the Northern Forum is understood by the Rovaniemi Code of Conduct (1994, p.4) as:

1) taking a long-term view which takes into consideration the interests of future generations, the co-existence of various industries and the health of local populations;
2) adherence to environmental standards which prevent pollution and the degradation of the land and resources;
3) accountability for environmental damage under which those who generate pollution pay for its consequences;
4) non-renewable resource development should be accomplished utilising best management practices to avoid damaging renewable resources; and
5) a precautionary principle to all development activities in the North, developing and using the best scientific information and technology available.

Norwegian initiative in 1993 started Barents Euro-Arctic cooperation. By developing the region, the Norwegians hoped to 'give it an eastern dimension by associating Murmansk and Archangel counties politically with the Barents cooperation' and 'a southern dimension by placing developments in the Barents region in a wider European framework wherever appropriate'. As the region is called the 'Euro-Arctic region,' it is 'part of a Nordic policy towards Europe which ties together this region and developments in Eastern and Southern Europe' (Stoltenberg, 1992, p.7).

For Norwegian Prime Minister Stoltenberg:

The Barents region will be an important part of the New Europe. It is richer in natural resources than any other region in Europe. The Barents Sea has Europe's richest fish stocks. There are extensive mineral deposits and productive forests in the area, and probably vast petroleum reserves, particularly in Russia (Stoltenberg, 1992, p.8).

The aim of Barents cooperation is to 'improve the conditions for local cooperation between local authorities, institutions, industry and commerce across the borders of a region' (Kirkenes Declaration, 1993). Collaboration on environmental issues is included in the Barents cooperation. The objective of the cooperation is to promote sustainable development in the region (see also Barents Euro-Arctic Council Environment Action Programme, 1994).

The interpretation of the idea of sustainable development has taken very concrete form in the Barents Euro-Arctic Cooperation; the members of the committee for the environment and economic cooperation had a joint session where environmental considerations were included into economic and investment projects. Another reason for the integration of economic and environmental concerns is that purely environmental projects would hardly be financed but funding could be found for an economic project involving the environment (Mähönen, 1997a).

Environmental cooperation in the 'Euro-Arctic Barents Region' has a European dimension, since two Arctic countries, Sweden and Finland, have become members of the European Union. Sweden has concentrated on presenting the Baltic region and its concerns within the European Union. Arctic affairs are therefore the responsibility of Finland. Finland became interested in developing the 'Northern Dimension of the European Union' in the fall of 1997. Though its EU membership Finland has 'the opportunity and the responsibility to introduce Arctic issues into EU fora and to promote the integration of Arctic environmental questions into the environmental policy of the EU' (Mähönen, 1997b, p.6).

The region is being mainly marketed to the European Union as a rich resource area. As the Finnish Prime Minister Paavo Lipponen stated at a recent Barents conference in Rovaniemi, Finland in September 1997: 'It is an area rich in resources, containing some of the world's largest reserves of natural gas and oil, resources of strategic importance to the [European] Union' (Lipponen, 1997, p.2). This region is described as 'sparsely populated and the climate is harsh'. Not only were the oil and gas reserves of the region emphasized but also the importance of timber resources and the forest industries: 'Russia's forests comprise over 20 per cent of the world's timber resources and more than half of the world's pine forests, giving a great potential to develop forestry and forest industries in Northwestern Russia'. The region has potential in terms of international trade: 'The increasing volume of trade and economic cooperation requires the development of infrastructure, particularly East-West rail and road connections' (Lipponen, 1997, p.2).

The fragility of the region is recognized: 'The guiding principles for all economic activities must be sustainable development. Arctic nature is highly vulnerable and has been exposed to serious pollution' (Lipponen, 1997, p.4).

The environmental problems are considered 'so extensive' that the political participation of groups outside of the Barents region is required. The Finnish view that, 'The opportunities of the European North are at the core of the evolving European Union's Northern Dimension' was how a recent Finnish report described the north (Nokkala, 1997, p.30).

Governmentality produces 'a mind' of its own. These process of discussing the mandate of the different Arctic fora produces its own rationality. The environmental discourse reflects the values and choice modern societies see in their relationship with nature. In the end, it is a question of the rationale for institutional functioning. These discussions are important and relevant since sustainable development is a thought or 'a state of mind' more than anything else (see Griffiths and Young, 1989, p.6).

By defining regimes differently the role of the cooperative arrangement in the Arctic can be appreciated. Regimes are not actors, all they do is constitute a specifically structured context within the interaction taking place (Rosenau, 1986, p.882). Instead of reifying a regime into a metaphysical agent, the identification of the respective regime as a discursively prestructured context allows us to appreciate and analyze the intersubjective processes that constitute the reality of the regime (Behnke, 1995, p.182). International society is understood as a notional: international society and its members, the sovereign states, are 'par excellence, notional beings, they exist in the minds of men and therefore, they actually exist objectively out there'. International society exists in the 'minds of many' and most importantly it exists in the 'minds of statesmen' (Wilson, 1989, p.53).

The different practices of cooperation in the Arctic make order in the region. As suggested by John Gerard Ruggie: 'The concept of regimes encompasses also the question how the order is achieved - regime is used to refer to common, deliberative, though often highly asymmetrical means of conducting interstate relations' (Ruggie, 1993a, p.12). Constructivism emphasizes the purposiveness of human action in international relations. Order is assumed to be desirable and disorder undesirable; where there is disorder, the aim of the restoration of order can also be identified. Order is never uncontentious. Attempts by one party to achieve order or address a disorder can often result in conflict and controversy with another. Not only is the conception of order contentious but there are also competing conceptions of order that might be mutually incompatible (McKinlay and Little, 1986, p.8-9).

Order is made by a community. For Norman Angell, who spoke of interdependence among the first in international relations, interdependence was 'a community of fates which is characterized by an intertwining of interests and mutual influence'(Wilde, 1991, p.89). The old institutionalist's view emphasizes the importance of a community among actors. The society, or community, is based on relationships and 'pull[s] people together in ways that are qualitatively different from the impersonal forces that create a system' (Brown, 1995, p.185; see also Waever, 1992). This view emphasizes the sense of community in international order; whatever order exists in a community it has a normative ground. For old institutionalists, order is a consequence of 'a sense of common interests in the elementary goals of social life' (Bull, 1977, p.65). E.H. Carr provides a deeply constitutive definition of a world community: 'There is a world community for the reason (and for no other), that people talk and within certain limits behave, as if there were a world community' (Carr, 1951, p.162).

This is, indeed, true for the idea of sustainable development; it is hardly more than talk. Doubts have been raised on whether these efforts of order making in the Arctic or anywhere else make any difference to the state of the environment. moment. The idea of sustainable development has become the corner stone of efforts to create a new international order. Yvonne Rydin asks: 'Can we talk ourselves into sustainability?' She questions whether through talk in different forums - local, regional and international - a new moral order based on ideas such as sustainable development can be created (Rydin, 1997). For example, according to Geoffrey Palmer: 'a strong argument can be made that during the times these instruments were being developed, the environmental situation in the world became worse and is deteriorating further' (quoted in Koskenniemi, 1994, p.43). This point suggests that developing regimes, institutions and mechanisms at the international level do not really have anything to do with the actual efforts to solve environmental problems. The problems are managed and controlled through international cooperation, but not solved. The order is maintained at least for the moment. The danger of preempting the idea of sustainable development in the negotiation process continues until today; everybody talks about but it has no content.

In the worst case, the tag 'sustainable development' can be added on whatever forms of cooperation are developed in the near future under the Arctic Council. Sustainability is accepted as such without questioning its basis. This is a real concern for the Arctic Council, too. The concept of sustainable development in 'empty' as Lynton Caldwell has noted comparing to the concept to 'a development horse... has pulled the cart of the economy, and the environment has been widely perceived as the paint on the cart'. The difficulty is that neither has a clearly defined destination (Caldwell, 1990b, p.179).

The Arctic and the Global

Governmentalities in the polar regions differ considerably. In 1991, international efforts to protect the polar regions were a focus of attention. From an early stage of cooperation the Arctic case was compared with the experience of cooperation in Antarctica. Finland, which made the initiative for a meeting on Arctic environmental protection, noted:

> We have had an interest in research activities in both Polar regions. Since 1984, we have paid increasing attention to the work in the Antarctic and supported measures aimed at promoting its sustainable development and in particular the preservation and conservation of its living resources (Bärlund, 1989, p.2).

The Arctic and Antarctic environment were both considered unique: 'Environmental disturbances in this cold climate [in the Arctic] pose more serious problems to the ecological balance than in other regions with the exception of Antarctica' (Statement of the Swedish Delegation, 1989, p.2). The development of contacts with the Antarctic Treaty System has brought the Netherlands and Chile into the work of the AEPS. The Antarctic Consultative Meeting in Seoul, in 1995, agreed to exchange of information on Arctic and Antarctic environmental issues (Chile, 1996).

The participation of non-Arctic states in Arctic cooperation has been based on their interest in polar regions; for example, Poland reaffirmed 'its interest in polar questions, both in Antarctic and in Arctica' by participating in the cooperation as an observer (Michiewisz, 1991, p.2).

However, the rationalities of governing in these regions are very different from each other. Article 2 of the Madrid Protocol designates the Antarctic as 'a natural reserve, devoted to peace and science'. The text in the Protocol continues:

> The protection of the Antarctic environment and dependent and associated ecosystems and the intrinsic value of Antarctica, including its wilderness and aesthetic values and its value as an area for the conduct of scientific research, in particular research essential to understanding the global environment, shall be fundamental considerations in the planning and conduct of all activities in the Antarctic Treaty area (In, Elliot, 1994, p. 223).

John Vogler points out the significant change that appears to have taken place in the negotiation of the Madrid Protocol can be seen as a rejection of the principle of managed resource exploitation in favor of a complete prohibition

on the 'wilderness and intrinsic values' of the Antarctic environment. However, this conclusion is not without reservations; according to Vogler, 'if there
was any immediate likelihood of economic exploitation of Antarctic mineral
resources, the situation would, in all probability, be rather different' (Vogler,
1996a, p.95). The governmentality of the Antarctic is one of valuing the
intrinsic merit of nature and the approach to the ecosystem. Through the
established Antarctic Treaty system, and developing the protocol on environmental protection, the Antarctic is constituted as region with a totally different
mentality to that of governing the Arctic.

At the UNCED in 1992 the Arctic was not mentioned nor included on the
agenda. In the preparations for the Nuuk ministerial meeting it was noted with
a disappointment: 'the meeting regretted the absence of an Arctic profile in the
draft Rio declaration' (Protecting the Arctic Environment, 1992, p.6). Raising
Arctic concerns at UN General Assembly Special Session (UNGASS), in New
York in June 1997, advanced the recognition of the global dimension of Arctic
problems. Arctic problems were raised by different states.

The connection between circumpolar and global developments was
emphasised. The Finnish minister of the Environment referred to 'the alarming
state of the Arctic environment'. The Finnish view was that the Arctic is 'a
sink of pollutants from industrial regions' because of the global circulation of
the atmosphere and the oceans. The most alarming risks are caused by organic
pollutants to human health (Haavisto, 1997b, p.2). The Norwegian prime
minister described the Arctic as 'one of the world's least polluted regions'.
However, the Norwegian view was that 'risks to this area are real, largely
caused by substances transported from sources outside the Arctic' (Jagland,
1997, p.2). The Canadian prime minister echoed these concerns: 'There is a
growing global consensus that environmental harm caused by some is a threat
to all' (Chretien, 1997, p.2).

According to Icelandic representative, 'climate change and increased
marine pollution threaten to have adverse effects and irreversible implications
worldwide'. The Icelandic view is that the threat of pollution to the Arctic is
real: 'Dangerous levels of pollution are accumulating in the Arctic, its origins in
diverse and remote parts of the place' (Oddson, 1997, p.4).

The report of the AMAP was presented in New York. The Danish prime
minister stressed the priority for the protection of the Arctic environment at
UNGASS (Rasmussen, 1997, p.2). 'The results of the AMAP program
provide a strong message to the rest of the world. Pollutants do not accept
borders...' according to the Danish views (Jensen, 1997, p.2). Spreading
information about the Arctic is considered important:

People in non-Arctic countries are seldom aware of the consequences of their decisions on the Arctic and how dependent they themselves actually are on a healthy Arctic environment (Haavisto, 1997a, p.1).

The UNGASS acknowledged that internationally a number of positive results were achieved in several social, economic and environmental components of sustainable development since Rio, in 1992. Some progress has been aimed at institutional development, international consensus-building, public participation, private sector action, and in curbing pollution and slowing the rate of resource degradation. However, the

...states are deeply concerned that the overall look for sustainable development is not much better today than what it was in 1992 in many parts of the world, especially in the least developed countries (UN Commission on Sustainable Development, 1997).

According to the participants, the general spirit at the Rio follow-up conference in New York in the summer of 1997 was disappointing. The progress five years after the Rio conference had not been as expected, nor were the results of the follow-up meeting as expected. Five years after UNCED the state of the global environment has continued to deteriorate and significant environmental problems remain deeply embedded in the socio-economic fabric of countries in all regions, according to the Report of the UN Commission on Sustainable Development (A/S-19/14).

At the Rio follow-up Arctic developments were, to some extent, recognized. The report on the adoption of Rio Declaration notes the Nuuk declaration made by the Arctic states in 1993 which acknowledges 'the special role of indigenous peoples in environmental management and development in the Arctic and of the significance of their knowledge and traditional practices, and will promote their effective participation in the achievement of sustainable development in the Arctic' (E/CN17/1997/8).Compared with other cooperative efforts in the region, the Arctic Council is the one that recognizes the role and status of indigenous peoples for sustainable development by granting them the status of 'permanent participants' (see The Earth Summit, 1993, p.415-416).

Regime Theory, Order and the Environment

Discourses and the Problem of Order

The discourses in cooperation are all efforts to make order in the Arctic. The problem of the environment is, above all, a problem of order. First, it is a problem since many environmental problems do not respect the borders of states. However, making order of problems is not only in the hands of the politicians, diplomats and government officials in international negotiations but these problems are a concern for many other actors as well. Second, it is a problem of making sense of the vast amount of information on the condition of the Arctic environment. It is a problem of ordering both the threats of today and the expectations of the future. Different reports and the work of researchers suggest the order of priorities. Finally, it is a problem of order in redefining the relations between states, other actors and institutions. It is a problem of order in the sense of finding a balance between different levels of international action.

Through constructing regimes, states make order or at least try to make order. International environmental regime-building depends on intersubjective meanings. As a phenomenon to be studied, regimes are experienced as having an existence beyond the individuals who happen to embody them at the time. Emphasizing the intersubjective quality of regimes underlines the fact that regimes are 'human artifacts' which have no existence in meaning apart from individuals or groups of human beings (Young, 1989, p.82).

The claim of this work is that human beings are enmeshed within webs of environmental relations and those relations are intersubjectively constituted. Therefore, the environment is a social construction. The environment is signified as 'whether a given aspect of social reality is a matter of contention or is regarded as natural and unproblematic, meaning is always imposed, not discovered, for the familiar world cannot be separated from the interpretative practices through which it is made' (Shapiro, 1989, p.11).

It has been a great challenge to include the ideas of Foucault into the regime analysis. The relationship between constructivist analysis and Foucaultian idea in this work has been complimentary; without the ideas of constructivist writers, there is little opportunity to use the Foucaultian ideas to study international regime building. Foucault's ideas on the other have provided a critical means to analyse the dynamics of regime-building.

Foucaultian order is not similar to the order of system theorists. For new institutionalists social order is seen as a consequence of instrumental relations among individual actors. Its individualism emphasizes the primacy of individual actors rather than of social collectives. Social order for a utilitarian is a derivative relation. Order derives entirely from equilibria in the instrumental relations and mutual expectations among rational egoistic individuals. This view suggests that whatever rules and regularities exist in the world it is a **product of an interplay of forces and lacking any kind of norm content; that** is, it is a system (Ashley, 1986, p.274-275).

For a constructivist, order is not a pattern: humans endow their behavior with purpose and meaning. It is essentially based on the idea of a community and on a shared set of rules, or as Foucault would prefer, on certain practices. One implication of the constructivist argument is that there is no conceptual difference between different levels of structure and order in the international system (see also Young, 1996, p.2-3). Following this idea, regimes are embedded in a fundamental institutional structure which constitutes the identity of the international society. As Nicholas Onuf points out, constructing levels of analysis is a practice of, 'ordering sets of relations, not as such, but as represented in our theories' (Onuf, 1995, p.52). For the constructivists, all institutions in the international system in fact order international life in both ways - structuring and ordering. Thus, the cooperative arrangements that the new institutionalists call regimes are not intrinsically different from the 'deep structural' institutions of the state system (Wendt and Duvall, 1989, p.63).

The Arctic effort at establishing order over a variety of environmental concerns and threats is a clear example of order making efforts in the international community. According to Foucault, regimes give order to a public space or realm of action. Foucault brings the idea of individualized regimes as 'localized power/knowledge' or 'localized orders' into the discussion. If a regime is a discourse/discipline set, it gives specific definition and order to a public space or realm of action (Keeley, 1990, p.96). Foucault directs our attention to patterns of understanding and organization which may not be shared by all but around which an order may be constructed.

Foucault's focus is on local orders and local struggles. He allows the treatment of regimes as efforts to organize a realm of action without overlooking fundamental contests. He looks at order-induced behavior and behavior that make sense only within the frameworks of a construction of reality. This construction of reality may affect, as well as reflect, networks of relations in a society (Keeley, 1990, p. 90).

Finally, this world is messy despite the continuous efforts of governing. The world is 'messy' because of the growing number of environmental problems or maybe more accurately because of a growing awareness of those environmental problems. This messiness, from the point of view of modernity, is temporary and eventually reparable; it can be replaced by the orderly and systematic rule of reason. Taking a more postmodern look means 'above all the tearing off the mask of illusions; the recognition of certain pretenses as false and certain objectives as neither attainable nor for that matter desirable'. A postmodern look on the messiness is that it will stay whatever we do or know. The little orders and systems that we carve out in the world are 'brittle, until-further-notice and as arbitrary and in the end contingent as their alternatives' (Bauman, 1993, p.33).

The Effects of Power

The standard view of regimes is that they are contracts when actors with long-term objectives seek to 'structure their relationships in a stable and mutually beneficial way' (Keohane, 1982, p.330). The contractarian view is based on a liberal approach which encourages regimes to be regarded as benevolent and voluntary associations (Keeley, 1990, p.84; see also Young, 1995b, p.189).The mainstream regime theory states that the problem facing an individual government is how to benefit from international exchange while maintaining as much autonomy as possible. Therefore, the problem in dealing with the commons is 'the management of interdependence in a system of sovereign states lacking the kinds of central authorities which are assumed to be providing order and regulation within domestic societies' (Vogler, 1996b, p.8). Many regime theorists focus on the observable political effects of institutions (Underdal 1995; see also Young and von Moltke, 1994; Underdal, 1992; Levy, 1993; Andresen and Wettestad, 1995; Bernauer, 1995). It appears for a realist that 'international collective self-regulation' denotes nothing more than the sum or regimes' contractual arrangements and international organizations (Behnke, 1995, p.189).

From a constructivist point of view regimes produce something more than just regulations for states. In emphasizing this aspect, international society is not only mediated and regulated by norms and rules but per se constituted by rules and norms which govern an actor's behavior. Studying these efforts of order-building challenges the view typical for realists and neorealists, international anarchy fosters competition and conflict between states and prevents cooperation even when they share common interests. Regimes are more than 'rare islands of cooperation in a sea of anarchy' possible under specific, circumscribed conditions (Behnke, 1995, p.187). They are also more than 'a small step removed from the underlying power capabilities that sustain them' (Krasner, 1982, p.191).

The rule of sovereignty is, of course, essential. Robert Keohane points out that the significance of international regimes does not lie in their formal legal status. The actions of sovereign states can overturn any practices of legal liability and property rights in international politics (Keohane, 1984, p.89). For 'new institutionalists', operational sovereignty within international institutions is distinguished from formal sovereignty. Operational sovereignty is understood as the legal freedom of the state to act under international law and formal sovereignty is defined in terms of the state's legal supremacy and independence. International institutional arrangements to deal with environmental problems, for example, constrain operational sovereignty but formal sovereignty remains largely intact (Levy, Keohane and Haas 1993, p.416-417; see also Krasner, 1989; Barkin and Cronin, 1994).

For 'old institutionalists', international institutions represent shared intersubjective understandings about the preconditions for meaningful state action. They constitute state actors as subjects of international life in the way as they make meaningful interaction by the latter possible. The old institutionalists were concerned with the ways in which international institutions structure those practices, the practices of state actors and their ordering (Wendt and Duvall, 1989, p. 53).

According to old institutionalists, human beings discover and express their individuality in communities and associations and practices which, for the most part, are not strictly contained within any specific state. International society is 'a social construction' (Dunne, 1995a and 1995b). International society exists when 'a group of states, conscious of certain common interests and common values, form a society in the sense that they conceive themselves to be bound by a common set of rules in their relations with one another, and share in the working of common institutions' (Bull, 1977, p.13).

Institutions in international society, according to old institutionalists, are 'recognized and established usages governing the relations between individuals or groups' (Wight, 1992, p.140-141). Their understanding of institutions is as a cluster of social rules, conventions, usages and practices. Institutions are a set of conventional assumptions held prevalently among the society members to provide a framework for identifying what the done thing is in appropriate circumstances (see Suganami, 1983, p.2365).

The rule of state sovereignty cannot be anything else but problematic: 'Politics is about the rule' (Ruggie, 1993b, p.151). The most important point in a recent debate on sovereignty was that state sovereignty is not a permanent principle of political life, because 'the appearance of permanence is simply an effect of complex practices working to affirm continuities and to shift disruptions and dangers to the margin', explains R.B.J Walker (Walker, 1993, p.163). The idea of continuous reconstitution of practices of power and sovereignty is repeated by Nicholas Onuf. He claims 'sovereignty is like an ideal that is never reached' (Onuf, 1989, p.142).

Instead of 'effectiveness' of cooperation, the concern here is the 'effects' of power in cooperation. Such an approach is not easily fitted in the expectations of 'effectiveness' of mainstream regime theories. Foucault suggests that power should be understood in terms of conduct; conduct is, simultaneously, to lead others and a way of behaving within a mostly open field of possibilities. Governing is the rationalization of the use of power. The exercise of power consists in guiding the possibility of conduct and putting the possible outcome in order.

The Question of S

It is, therefore, interesting to study how sovereignty over the environment is constituted and reconstituted - even in the circumpolar region. A recent Canadian government report notes that 'on the question of sovereignty, the Arctic Council cannot help to resolve long-standing disputes between members'. Nevertheless, the report expects that over time such disputes are likely to become less important as regional cooperation replaces national sovereignty as a priority for the Arctic states. (Report of the House of Commons Standing Committee on Foreign Affairs and International Trade, 1997, p.81).

In speaking of sovereignty as domination, Foucault does not have in mind the solid and global kind of domination that one person exercises over others, or one group over another but the manifold forms of domination that can be exercised within society (Foucault, 1980a, p.96). Regarding the relations between states and indigenous peoples, the Foucaultian approach means that sovereignty should be carefully studied as a theory and practice of power. This line of thought seeks to question conventional political thought 'couched in terms of sovereignty to understand the complexity and diversity of modern power'. Foucault suggests that the focus should not be the juridical edifice of sovereignty, the state apparatuses and the ideologies which accompany them (Hunt and Wickham, 1994, p.44).

The point is not to study the 'dark side of law' but disciplinary mechanisms of power. Therefore sovereignty, neither as a form of law nor domination, should be taken as granted but looked at closely as how it is maintained by a particular set of practices by states. However, from a Foucaultian point of view, this collective existence is of domination. Foucault's view is that rules and rights in particular should be viewed not in terms of legitimacy to be established but in terms of the methods of subjugation that it instigates (Foucault, 1980a, p.95-96).

As Alan James notes 'states may not be verbally explicit about the nature of their sovereignty ...But their actions make the meaning which they attach to the term sovereignty entirely plain' (James, 1986, p.22). The actions of the Arctic Council in handling the environmental challenge advances the importance and role of the states. National responsibilities for protecting the Arctic environment are emphasized by the AEPS. According to the Canadian representative, it is clear that 'circumpolar cooperation, bilaterally and multilaterally, flows naturally from our domestic priorities'. International cooperation is connected to national or domestic concerns:

> We [Canadians] did not adopt a policy of promoting cooperation merely because it seemed like an interesting idea. We adopted it because we saw that increased cooperation with our circumpolar neighbors would help us meet the many practical economic, environmental, social and cultural challenges we face as an Arctic nation (Campeau, 1990, p. 4).

The AEPS was meant to be implemented 'through national legislation and in accordance with international law' (AEPS, 1991, p.3). The national responsibilities of each state were emphasized in Nuuk; according to the ministers, 'effective domestic environmental legislation is a prerequisite to the protection of the environment'. The ministers committed themselves to promote the legislation required for the conservation of the Arctic environment (The Nuuk Declaration on Environment and Development in the Arctic, 1993).

The importance of national measures was stressed by the AMAP. According to the AMAP report:

> Arctic countries should take all necessary steps to ensure that their domestic responsibilities and arrangements to reduce contaminant inputs to the Arctic region are fully implemented (AMAP, 1997, p.xii).

If these responsibilities are not addressed, 'the justification for recommending actions aimed at reducing transboundary contaminants with sources outside of the Arctic will be accordingly diminished' (AMAP, 1997, p.xii).

The practice of state sovereignty is the main concern here: As Carl Hild (1996) from Alaska explained, 'the AEPS is cooperation, government to governments; indigenous peoples' organizations are allowed to observe the negotiations'. A discussion on the relations of sovereignty, self-determination and environmental concerns are unavoidable in the AEPS and trying to make the Arctic Council work. The states, in particularly the United States, prefer clear limits in the role of the IPO's in the work of the Arctic Council. A good example of this is the problem with using the word 'peoples' which emerged in the fall 1995 Arctic Council negotiations. The United States suggested that instead of using the words 'indigenous peoples' (with s) the words 'indigenous people' should be used in the declaration text (Arctic Council Declaration Draft, October 20, 1995). The U.S. representatives pointed out that 'Indigenous Peoples' should not be capitalized throughout the text since it was not a proper name. The use of 's' on peoples was questioned and the U.S. recommended that it be dropped (Arctic Council Declaration Draft, October 20, 1995).

Through this rather minor issue of one letter, Arctic environmental cooperation cannot be separated from the efforts of international society to advance the rights of indigenous peoples. The 1989 ILO Convention speaks of indigenous peoples thus: 'the use of the term ´peoples´ in this convention shall not be construed as having any implications as regards the rights which may attach to the term under international law'. However, this note in the ILO Convention did not prevent the Arctic Council papers including a note about the use of the word 'peoples'. This note stresses that use of the word 'peoples' should not be understood to refer to the rights of peoples according to international practice. This note is now added practically everywhere in the AEPS papers, mainly because of the demand by the US representatives. In the final version of the declaration establishing the Arctic Council there was a note attached to the text stating that 'the use of the term ´peoples´ in this declaration shall not be construed as having any implications as regard the rights which may attach to the term under international law' (Declaration on the Establishment of the Arctic Council, 1996).

Unless one considers the issue of sovereignty, the concern of the U.S. representatives is difficult to understand. The interesting point is how the issues of sovereignty, self-determination and governmentality are embedded into Arctic environmental cooperation.With the AEPS and the Arctic Council, the rule of sovereignty does indeed collapse into practices of power. The success in the Arctic is in the hands of the states. Studying the discourse on sovereignty suggests considering the cooperation as an event in the use of power, both in its restrictive and productive sense, in defining the relationship between the state and its indigenous peoples.

In the Foucaultian sense, regimes as social institutions work within the thin but entangling webs of power relations. Power relations are not external to a regime. Foucault's view is that power is always present in social arrangements and it is always exercised. Institutions are 'the most readily definable macro-objects, grosser instruments for the finer, more elemental workings of power'. Institutions are relevant in terms of places where power 'becomes embodied in techniques and equips itself with instruments and eventually even violent means of material intervention' (Caputo and Yount, 1993, p.9). Foucault suggests studying geographical dimensions as, 'tactics and strategies of power which deployed through implantations, distributions, demarcations, control of territories and organizations of domains...' form a region (Foucault, 1980c, p.77).

The logic of governmentality makes indigenous peoples participants in the cooperation with the states and users of the resources in the region, but not owners of these resources. The issues of sovereignty, self-determination and governmentality are embedded into Arctic environmental cooperation making states to emphasise national responsibilities and state sovereignty in the region. With the AEPS and the Arctic Council, the rule of sovereignty does collapses into practices of power. More important than finding solutions to the problems at hand was to secure existing power relations both inside and outside individual Arctic states, including avoiding the problem of dealing with the demands of self-government and land ownership increasing among different groups of indigenous peoples in the Arctic.

Telling the story of the Arctic cooperation including the views of indigenous peoples has not been an easy task. As far I understand, the task of an intellectual is not to shape the political will of another nor take sides on behalf of anybody else. Indigenous peoples have to fight their own struggles. However, I'm aware that my actions as a researcher may have consequences, even political ones. It is possible that my actions make no difference at all. I have, however, tried to follow the logic suggested by Foucault:

> ... to question over and over again what is postulated as self-evident, to disturb people's mental habits, the way they do and think things, to dissipate what is familiar and accepted, to reexamine rules and institutions... to participate in the formation of a political will (Foucault, 1988b, p.xvi).

I do hope that I have managed to shake some of 'mental habits' created in the Arctic cooperation so far as well as some of those in my own field - international relations.

Bibliography

Unpublished Material and Documents

Address of the Soviet Representative (Unofficial translation) (1989), in *Consultative Meeting on the Protection of the Arctic Environment*, Annex II:1-12 and Annex III:1-25 of the Final Report, Rovaniemi, Finland, September 20-26, 989, Helsinki.

AEPS (1991), *Arctic Environmental Protection Strategy*, Rovaniemi, Finland.

AEPS (1995), *Procedures for Observer Participation and Accreditation* (Draft). SAAO Meeting, Toronto, Canada, November 29 - December 1, 1995.

Aikio, P. (1990), 'The Circumpolar Peoples and the Protection of the Arctic Environment. A Saami Viewpoint', in *Protecting the Arctic Environment. Report on the Yellowknife Preparatory Meeting*, Annex II:18, Yellowknife, NWT, Canada, April 18-23, 1990, Ottawa.

Alta Declaration on the Arctic Environmental Protection Strategy (1997), Alta, Norway, June 12-13, 1997, http://arctic-council/usgs.gov, 26.8.1999.

Arctic Council Declaration Draft, June 20, 1995.

Arctic Council Declaration Draft, August 16, 1995.

Arctic Council Declaration Draft, October 20, 1995.

Arctic Council Declaration Draft, January 15, 1996.

Arctic Council Declaration Draft, April 19, 1996.

Arctic Council Declaration Draft, June 9, 1996.

Arctic Council SAO (1997), in *Arctic Council Meeting of SAOs and Permanent Participants. Information Kit*, Ottawa, Canada, October 7-9, 1997. The Arctic Environment (1993), *Report: Second Ministerial Conference 16 September 1993 - Nuuk Greenland*, Ministry of Foreign Affairs, Copenhagen.

A/S-19/14, *Report of the Commission on Sustainable Development on Preparations for the Special Session of the General Assembly for the Purpose of an Overall Review and Appraisal of the Implementation of Agenda 21*, gopher:// gopher.un.org:70/00/ga/s-19/plenary/AS19--14.EN. 25.6.1997.

Barents Euro-Arctic Council Environment Action Programme (1994), Bodø, Norway, June 15, 1994.

Bärlund, K. (1989), Opening Statement, in *Consultative Meeting on the Protection of the Arctic Environment*, Annex II:1-12 and Annex III:1-25 of the Final Report, Rovaniemi, Finland, September 20-26, 1989, Helsinki.

Berntsen, T. (1997), Statement, in AEPS Ministerial Meeting, Alta, Norway, June 12-13, 19997 in *Arctic Council Meeting of SAOs and Permanent Participants. Information Kit*, Ottawa, Canada, October 7-9, 1997.

Bjarnason, G. (1997), Opening Statement, in *AEPS Ministerial Meeting*, Alta, Norway, June 12 -13, 1997.

Brelsford, T. (1995), *A Compilation and Summary of Ethical Principles for Arctic Research*, CAFF Working Group, For Consideration by the SAAOs, Iqaluit, Canada, March 15-17, 1995.

Brelsford, T. (1996b), *Status Report on Development of Ethical Principles for Arctic Research*. Presented to the CAFF Working Group Meeting, Rovaniemi, Finland, September 9-12, 1996.

CAFF (1994), *The State of the Protected Areas*, CAFF.

CAFF (1995), *Paper on Integration of Indigenous Knowledge*.

CAFF (1995-1996), *Annual Workplan 1995-1996*.

CAFF (1996), *Co-Operative Implementation Strategy for the Convention of Biological Diversity in the Arctic Region*, Draft February 1996.

CAFF (1997a), *Habitat Conservation Report*, Report No. 2, http://www.grida.no/caff/hcr2.htm, 25.6.1997.

CAFF (1997b), *Threats to Arctic Species and Ecosystems*, http://www.grida.no/caff/threats.htm, 25.6.1997.

Campeau, A. (1990), 'Notes by Special Advisor on International Affairs to the Minister of the Environment, on behalf of the Honourable Lucien Bouchard, Minister of the Environment, to the Yellowknife Preparatory Meeting on the Protection of the Arctic Environment', in *Protecting the Arctic Environment. Report on the Yellowknife Preparatory Meeting*, Annex III:12, Yellowknife, NWT, Canada, April 18-23, 1990, Ottawa.

Chile (1996), Statement, at the *Third AEPS Ministerial Meeting*, Inuvik, Northwest Territories, Canada, March 20-21, 1996.

Chretien, J. (1997), *Notes for an Address on the Occasion of the United Nations General Assembly Special Session on Sustainable Development*, 24 June 1997, gopher://gopher.un.org:70/00/docs/S-19/statements/gov/ CHRETIEN. TXT.

Consultative Meeting on the Protection of the Arctic Environment (1989), Report and Annex 1:1-7, Rovaniemi, Finland, September 20-26, 1989, Helsinki.

Dahl, B. (1991), Statement, in *Ministerial Conference on the Protection of the Arctic Environment*, Rovaniemi, Finland, June 13-14, 1991.

Danilov-Danilian, V.I. (1993), The Address of the Russian Delegation Head at the *Second Ministerial Meeting of the Arctic Countries*, Nuuk, Greenland, September 16, 1993.

Declaration on the Establishment of the Arctic Council (1996), Ottawa, Canada, September 19, 1996, http://arctic-council/usgs.gov/, 26.8.1999.

Declaration on the Protection of the Arctic Environment (1991), Rovaniemi,

Finland, June 14, 1991, http://arctic-council/usgs.gov/, 26.8.1999.

E/CN.4/Sub.2/1991/8, Human Rights and the Environment. Preliminary Report by Mrs. Fatma Zohra Ksentini, Special Rapporteur, pursuant to Sub-Commission Resolutions 1990/7 and 1990/27, in *Commission on Human Rights, Sub-Commission on Prevention of Discrimination and Protection of Minorities*, 43rd Session.

E/CN17/1997/8, *Rio Declaration on Environment and Development: Application and Implementation. Report of the Secretary General,* gopher://gopher.un.org:70/00/esc/cn17/1997/off/97--8.EN, 25.6.1997.

EPPR (1996), *Status Report For the Period between the Ministerial Meeting in Nuuk, Greenland 16 September 1993 and the Ministerial Meeting in Inuvik, Canada, 20-21 March 1996.*

Eriksen, S. (1997), Letter to Mary Simon, Ambassador for Circumpolar Affairs, September 18, 1996, in *Arctic Council Meeting of SAOs and Permanent Participants, Information Kit*, October 7-9, 1997 Ottawa, Canada.

Final Draft WG1 Arctic Environmental Protection Strategy (1991), in *Protecting the Arctic Environment*, Kiruna Preparatory Meeting, January 15-17, 1991.

Gudnason, E. (1991), Statement, in *Ministerial Conference on the Protection of the Arctic Environment*, Rovaniemi, Finland, June 13-14, 1991.

Haarder, B. (1991), Statement, in *Ministerial Conference on the Protection of the Arctic Environment*, Rovaniemi, Finland, June 13-14, 1991.

Haavisto, P. (1997a), Opening Statement, at the *AEPS Ministerial Meeting*, Alta, Norway, June 12-13, 1997.

Haavisto, P. (1997b), Statement, at the *Special Session of the UN General Assembly for the Overall Review and Appraisal of the Implementation of Agenda 21.* New York, June 25, 1997, gopher://gopher.un.org:70/00/docs/S-19/statements/gov/HAAVISTO.TXT.

Halonen, L. (1991), Statement, in *Ministerial Conference on the Protection of the Arctic Environment*, Rovaniemi, Finland, June 13-14, 1991.

Halonen, L. (1996b), *Arktisk miljö- og annet samarbeid*, in Samenes 16. Konferens og Barents Urfolkskonferense, 15-18.10.1996 i Murmansk.

Henriksen, John B. (1996), *The Legal Status of Saami Land Rights in Finland, Russia, Norway and Sweden*, Sami Council.

Huntington, H. (1997), *Presentation at the Conference on Environmental Pollution in the Arctic*, Tromsø, Norway, June 4, 1997.

ICC (1990), 'Proposed Objectives for an Arctic Sustainable and Equitable Development Strategy. Submitted by Mary Simon, President', in *Protecting the Arctic Environment. Report on the Yellowknife Preparatory Meeting*, Annex II:15, Yellowknife, NWT, Canada, April 18-23, 1990, Ottawa.

ICC (1995a), *Report to the Working Group on Emergency Prevention, Preparedness and Response for the Norilsk Meeting*, August 1995.

ICC (1995b), *Collapse of the Arctic Seal Skin Market: Retrospective Study and*

Sustainable Options, Prepared for the Arctic Environmental Protection Strategy Task Force on Sustainable Development and Utilization.

Iceland (1990), Statement, in *Protecting the Arctic Environment. Report on the Yellowknife Preparatory Meeting*, Annex III:8, Yellowknife, NWT, Canada, April 18-23, 1990, Ottawa.

Introductory Statement by Assistant Under-Secretary, Desiree Edmar (1990), in *Protecting the Arctic Environment. Report on the Yellowknife Preparatory Meeting*, Annex III:6. Yellowknife, NWT, Canada, April 18-23, 1990, Ottawa.

Inuvik Declaration on Environmental Protection and Sustainable Development in the Arctic (1996), Inuvik, Northwest Territories, Canada, March 21, 1996, http://arctic-council/usgs.gov, 26.8.1999.

Jaakkola, E. (1989), 'Conservation of Arctic Living Resources', in *Consultative Meeting on the Protection of the Arctic Environment*, Annex II:1-12 and Annex III:1-25 of the Final Report, Rovaniemi, Finland, September 20-26, 1989, Helsinki.

Jagland, T. (1997), Statement, at the *General Assembly 19th Special Session on the Review and Appraisal of the Implementation of Agenda 21*. New York, June 23,1997, gopher://gopher.un.org:70/00/docs/S-19/statements/gov/jag land.txt.

Jensen, M. (1997), Opening Statement, at the *AEPS Ministerial Meeting*, Alta, Norway, June 12-13, 1997.

Kirk, E.J. (1997), *Scientific and Technical Cooperation in the Russian North West*. Paper presented at the 37th Annual Convention of the International Studies Association, Toronto, Canada, March 18-20,1997.

Kirkenes Declaration (1993), Cooperation in the Barents Euro-Arctic Region, Conference of Foreign Ministers in Kirkenes, Norway, 11 January 1993.

Kjellen, B. (1997), Statement in the AEPS Ministerial Meeting, Alta Norway, June 12-13, 1997, in *Arctic Council Meeting of SAOs and Permanent Participants. Information Kit*, October 7-9, 1997 Ottawa, Canada.

Kuokkanen, R. (1997), 'Statement by Sami Council', in *Commission on Human Rights. Subcommission on Prevention of Discrimination and Preventionof Minorities, Working Group on Indigenous Populations*, Fifteenth Session 28 July-1 August 1997, Item 5 (a) Environment, land and sustainable development.

Kuptana, R. (1996), Opening Remarks, in *Third AEPS Ministerial Meeting*, Inuvik, Northwest Territories, Canada, March 20-21, 1996.

Lipponen, P. (1997), *The European Union Needs a Policy for the Northern Dimension* in Barents Region Today Conference, Rovaniemi, Finland, September 15, 1997.

Lock, G.S.H. (1995), *Arctic Science Policy and Advice*. Presented at the International Conference for Arctic Research Planning, sponsored by the International Arctic Science Committee and the Polar Research Board of the National Academy of Sciences and the National Science Foundation of the United States. Dartmouth College, Hanover, USA, December 5-9, 1995.

Lynge, A. (1991), Statement, in *Ministerial Conference on the Protection of the Arctic Environment*, Rovaniemi, Finland, June 13-14, 1991.

Lynge, A. (1997), Statement in AEPS Ministerial Meeting, Alta, Norway, June 12-13, 1997,in *Arctic Council Meeting of SAOs and Permanent Participants. Information Kit*, October 7-9, 1997 Ottawa, Canada.

Meeting of Working Group on Permanent Participants (1995), Arctic Council Discussions, November 26, 1995, Toronto.

Michiewisz, J. (1991), Statement, in *Ministerial Conference on the Protection of the Arctic Environment*, Rovaniemi, Finland, June 13-14, 1991.

Nurmi, M. (1996), Address by Head of the Finnish Delegation, at *Third AEPS Ministerial Meeting*, Inuvik, Northwest Territories, Canada, March 20-21, 1996.

The Nuuk Declaration on Environment and Development on the Arctic (1993), Nuuk, Greenland, 16 September, 1993.

Oddson, D. (1997), Statement, at the UN 19th Special Session., in *UN Press Release* GA/9264, ENV/DEV/430, June 24, 1997.

Olsen, O.R. (1991), Speech, in *Ministerial Conference on the Protection of the Arctic Environment*, Rovaniemi, Finland, June 13-14, 1991.

Opening Statement by Head of the Norwegian Delegation, Mr. Dagfinn Steneth (1990), in *Protecting the Arctic Environment. Report on the Yellowknife Preparatory Meeting*. Annex III:2. Yellowknife, NWT, Canada, April 18-23, 1990, Ottawa.

Petersen, T. (1997), *Presentation at the Conference on Environmental Pollution in the Arctic*. Tromsø, Norway, June 4, 1997.

Pietikäinen, S. (1991), Statement, in *Ministerial Meeting on the Protection of the Arctic Environment*, Rovaniemi, Finland, June 13-14, 1991.

Prokosch, P. (1997), 'An Application to Arctic Council for Observer Status, WWF, July 23, 1997', in *Arctic Council Meeting of SAOs and Permanent Participants. Information Kit*, October 7-9, 1997 Ottawa, Canada.

Proposal by the Government of Canada (1990), in *Protecting the Arctic Environment. Report on the Yellowknife Preparatory Meeting*, Yellowknife, NWT, Canada, April 18-23, 1990, Ottawa.

Protecting the Arctic Environment (1990), Report on the Yellowknife Preparatory Meeting. Yellowknife, NWT, Canada, April 18-23, 1990, Ottawa.

Protecting the Arctic Environment (1991), Kiruna Preparatory Meeting, January 15-17, 1991.

Protecting the Arctic Environment (1992), Summary Report of a Meeting in Copenhagen, Denmark, April 23-24, 1992.

Rasmussen, P.N. (1997), Statement, at the *UN 19th Special Session of the General Assembly (UNGASS) 23-27 June, 1997*, New York, June 23, 1997, gopher://

gopher.un.org/00/docs/S-19/statements/gov/RASMUSSEN.TXT.

Reimer, C. (1996), *Financing Indigenous Peoples' Participation in the Arctic Council. A Discussion Paper for Presentation to the AEPS and Arctic Council Senior Arctic Officials*, November 1996.

Report of the CAFF Chair and Secretariat (1996), in *Meeting of the Senior Arctic Affairs Officials and Third Ministerial Conference*, Inuvik, Northwest Territories, Canada, March 18-21, 1996.

Report of the Third Ministerial Conference on the Protection of the Arctic Environment (1996), Inuvik, Northwest Territories, Canada, March 20-21, 1996.

Roots, F. (1991), Remarks, in *Ministerial Conference on Protection of the Arctic Environment*, Rovaniemi, Finland, June 13-14, 1991.

Rovaniemi Code of Conduct (1994), Northern Forum.

Royal Commission on Aboriginal Peoples (1996), *Highlights from the Report of the Royal Commission on Aboriginal Peoples. People to People, Nation to Nation*, http://www.inac.gc.ca/rcap/report/index.html, 10.6.1997.

Rydin, Y. (1997), *Can We Talk Ourselves into Sustainability?* Plenary Speech at the Research Conference 'The Environment, Society and Sustainability - A Nordic Perspective', August 25-27, 1997, in Oslo, Norway.

SAAO (1995), *Final Minutes*, March 15-17, Iqaluit, Northwest Territories, Canada.

SAAO (1997), *SAAO Report to the Ministers for the Fourth Ministerial Conference on the Arctic Environmental Protection Strategy (AEPS)*, Alta Norway, June 12-13, 1997.

Sámi Programme of the Environment (1990 [1986]), in *Protecting the Arctic Environment. Report on the Yellowknife Preparatory Meeting*, Annex II:17, Yellowknife, NWT, Canada, April 18-23, 1990, Ottawa.

Scully, T. (1997), Statement in AEPS Ministerial Meeting, June 12-13, 1997, in *Arctic Council Meeting of SAOs and Permanent Participants. Information Kit*, October 7-9, 1997 Ottawa, Canada.

Second Conference of Parliamentarians of the Arctic Region (1996), *Conference Statement*, Yellowknife, Canada, March 14, 1996, http://www.grida.no/ parl/-conf/index.htm, 26.8.1999.

Siddon, T. (1991), Notes for Remarks, in *Ministerial Conference on the Protection of the Arctic Environment*, Rovaniemi, Finland, June 13-14, 1991.

Simon, M. (1990), 'Towards an Arctic Sustainable and Equitable Development Strategy; Some Preliminary Views', in *Protecting the Arctic Environment. Report on the Yellowknife Preparatory Meeting*, Yellowknife, NWT, Canada, April 18-23, 1990, Ottawa.

Simon, M. (1995), Keynote Speech, in *International Conference for Arctic Research Planning*', Hanover, NH, 5-9 December, 1995.

Simon, M. (1997a), Speaking Notes for Opening Statement, at the *AEPS Ministerial Conference*, Alta, Norway, June 12, 1997.

Simon, M. (1997b), Closing Remarks, at the *AEPS Ministerial Conference*, Alta, Norway, June 13, 1997.

Solovyanov, A. (1997), Statement, in AEPS Ministerial Meeting, Alta, Norway, June 12-13, 1997 in *Arctic Council Meeting of SAOs and Permanent Participants. Information Kit*, October 7-9, 1997 Ottawa, Canada.

Statement by the Canadian Delegation (1989), in *Consultative Meeting on the Protection of the Arctic Environment*, Annex II:1-12 and Annex III:1-25 of the Final Report, Rovaniemi, Finland, September 20-26, 1989, Helsinki.

Statement by the Chair (1990), in *Protecting the Arctic Environment. Report on the Yellowknife Preparatory Meeting*, Annex II:5, Yellowknife, NWT, Canada, April 18-23, 1990. Ottawa.

Statement by the Danish Delegation (1989), in *Consultative Meeting on the Protection of the Arctic Environment*, Annex II:1-12 and Annex III:1-25 of the Final Report, Rovaniemi, Finland, September 20-26, 1989, Helsinki.

Statement by the Finnish Delegation (1989), in *Consultative Meeting on the Protection of the Arctic Environment*, Annex II:1-12 and Annex III:1-25 of the Final Report, Rovaniemi, Finland, September 20-26, 1989, Helsinki.

Statement by the Swedish Delegation (1989), in *Consultative Meeting on the Protection of the Arctic Environment*, Annex II:1-12 and Annex III:1-25 of the Final Report, Rovaniemi, Finland, September 20-26, 1989, Helsinki.

Statement from the Soviet Delegation (1990), in *Protecting the Arctic Environment. Report on the Yellowknife Preparatory Meeting*, Annex III:7, Yellowknife, NWT, Canada, April 18-23, 1990, Ottawa.

Statement of ECE (1989), in *Consultative Meeting on the Protection of the Arctic Environment*, Annex II:1-12 and Annex III:1-25 of the Final Report, Rovaniemi, Finland, September 20-26, 1989, Helsinki.

Statement of the Norwegian Delegation (1989), in *Consultative Meeting on the Protection of the Arctic Environment*, Annex II:1-12 and Annex III:1-25 of the Final Report, Rovaniemi, Finland, September 20-26, 1989, Helsinki.

Stoltenberg, J. (1991), Statement, in *Ministerial Conference on the Protection of the Arctic Environment*, Rovaniemi, Finland, June 13-14, 1991.

Stone, D. (1997), Opening Speech by the AMAP Working Group Chairman at *the Conference on Environmental Pollution in the Arctic*, Tromsø, Norway, June 1-5, 1997.

TFSDU (1994), *Terms of Reference and Workplan.*

TFSDU (1995a), *Summary of the Meeting of the Task Force on Sustainable Development and Utilization*, March 13-14, 1995.

TFSDU (1995b), *Regional Application of Agenda 21 in the Arctic* (Draft), November 1995.

TFSDU (1996), *Review of Work in Progress.*

Towards an Inuit Regional Conservation Strategy (1986), ICC Environmental Commission.

Trade Barriers Affecting Products from the Arctic (1995), Discussion Paper.

Tromsø Declaration (1993), Northern Forum.

UN Commission on Sustainable Development, Revised Draft Political Statement (1997), proposed by the Mostafa K. Tolba and Monica Linn Locher, gopher://gopher.un.org:70/00/ga/docs/S-19/plenary/DRAFTS, 25.6.1997.

Utsi, N. (1996), Statement on behalf of the Saami Council, in *Third AEPS Ministerial Conference*, Inuvik, Northwest Territories, Canada, March 20-21, 1996.

Weinman, J.G. (1991), Remarks, in *Ministerial Conference on the Protection of the Arctic Environment*, Rovaniemi, Finland, June 13-14, 1991.

WWF (1995), *Comments to the Declaration on the Establishment of the Arctic Council*.

WWF (1997), Statement, at the *AEPS Ministerial Meeting*, Alta, Norway, June 12-13, 1997. Young, O.R. (1993a), *Regions in International Society*, Mimeo.

Young, O.R (1995a), *The Arctic Environmental Protection Strategy: Looking Back ward, Looking Forward*, An Essay Based on Presentations Made at the Annual Danish/American Greenland Science Conference Held in Copenhagen, Denmark on 7 April 1995 and at the Session of the 1995 Calotte Academy Held in Inari, Finland on 21 May 1995.

Young, O.R. (1996), *Arctic Governance: Meeting Challenges of Cooperation in the High Latitudes*, An Essay Prepared for Discussion at the Second Conference of Parliamentarians of the Arctic Region, 13-14 March 1996 Yellowknife, Northwest Territories, Canada.

Young, O. (1997), International Arctic Science Committee (IASC) Comments, in *SAAOs/SAOs Meeting*, 9-10 June 1997.

Interviews and Personal Communication

Brelsford, T. (1996a), U.S. Fish and Wildlife Service, Anchorage, Interview, February 1996.

Clark, J. (1996), Northern Forum Secretariat, Anchorage, Interview, February 1996.

Cline, D. (1996), National Audubon Society, Anchorage, Interview, February 1996.

Cochran, P. (1996), Alaska Native Science Commission, Anchorage, Interview, February 1996.

Elling, H. (1996), Danish Polar Centre, Copenhagen, Personal communication, June 1996.

Fenge, T. (1995), Canadian Arctic Resources Committee, Interview (by phone), October 1995.

Gibson, M. (1995), Arctic Network, Anchorage, Interview (by phone), October 1995.

Gibson, M. (1996), Arctic Network, Anchorage, Interview, February 1996.

Halonen, L. (1996a), Sami Representative to AEPS, Interview, August 1996.

Hild, C. (1996), Rural Cap, Anchorage, Interview, February 1996.

Hilden, M. (1996), Arctic EIA Project, Finnish Environment Agency, Helsinki, Interview, March 1996.

Huntington, H. (1995), ICC, Environmental Coordinator, Interview (by phone), October 1995.

Huntington, H. (1996), ICC, Environmental Coordinator, Interview, February 1996.

Hurst, R. (1996), AEPS Secretariat, Environment Canada, Hull, Interview, January 1996.

Hurwich, E. (1995), Circumpolar Conservation Union, Interview (by phone), October 1995.

Kankaanpää, P. (1996), Finnish AEPS Coordinator, Ministry of the Environment, Helsinki, Inteview, October 1996.

Lassig, J. (1996), Finnish Representative to PAME, Ministry of the Environment, Helsinki, Interview, March 1996.

Lekanof, F. (1996), Aleutian/Pribilof Islands Association, Environmental Program Coordinator, Anchorage, Interview, February 1996.

Liljelund, L-E. (1996), AMAP Assessment Working Group Chair, Ministry of the Environment, Stockholm, Interview, June 1996.

Lindström, G. (1996), Executive Secretary, Nordic Council of Ministers, Helsinki, Interview, March 1996.

McCloskey, M. (1995), Sierra Club, Interview (by phone), October 1995.

Mähönen, O. (1996), Finnish AEPS Coordinator, Ministry of the Environment, Helsinki, Interview, May 1996.

Mähönen, O. (1997a), Finnish AEPS Coordinator, Environment Agency, Rovaniemi Personal communication, June 1997.

Matero, S. (1997), Sami Programme on Sustainable Development, Finnish Sami Parliament, Personal communication, April 1997.

Mayo, R. (1996), Council of Athabascan Tribal Governments, Fairbanks, Interview, February 1996.

Mills, C. (1995), Dene Nation, Personal communication, December 1995.

Nenonen M.-L. (1996), Finnish AMAP Representative, Environment Agency, Rovaniemi, Interview, May 1996.

Pahkala, O. (1996), Finnish representative in the EPPR, Ministry of the Environment, Helsinki, Interview, March 1996.

Pagnan, J. (1996), CAFF Secretariat, Environment Canada, Hull, Interview, January 1996.

Petersen, T. (1996), Indigenous Peoples Secretariat, Acting Executive Secretary, Copenhagen, Interview, June 1996.

Prokosch, P. (1996), WWF Arctic Program, Oslo, Personal communication, June 1996.

Puurunen, H. (1996), Finnish SAAO, Ministry of Foreign Affairs, Helsinki, Interview, April 1996.

Reimer, C. (1995), ICC, Research Director, Interview (by phone), October 1995.
Reiersen, L.-O. (1996), Executive Director, AMAP Secretariat, Oslo, Interview, June 1996.
Rouhinen, S. (1996), Finnish Representative to TFSDU, Ministry of the Environment, Helsinki,Interview, March 1996.
Senseney, R. (1996), U.S. SAAO, State Department, Washington DC, Interview, February 1996.
Simon, M. (1996), Canadian Arctic Ambassador, Ottawa, Interview, January 1996.
Skåre, M. (1996), Ministry of Foreign Affairs, Legal Adviser, Oslo, Interview, June 1996.
Snider, A. (1996), AEPS Secretariat, Environment Canada, Hull, Interview, January 1996.
Sondergård, J. (1996), Greenlandic Home Rule Office, Copenhagen, Interview, June 1996.
Torikka, R. (1996), Sami Representative to CAFF, Interview, April 1996.
Whitby, L. (1996), Canadian SAAO and Chair of the TFSDU, Environment Canada, Ottawa, Interview, January 1996.
Wohl, P. (1996), Northern Forum Secretariat, Environmental Health and Emergency Response, Anchorage, Interview, February 1996.

Published Documents and Literature

A/CONF.48/14/Rev.1, *Report of the United Nations Conference on the Human Environment (1972)*, Stockholm, Sweden, 5-16 June 1972.
Adler, E. and Haas, P.M. (1992), 'Conclusion: Epistemic Communities, World Order, and the Creation of a Reflective Research Program', *International Organization*, vol. 46, 1, pp.367-390.
Alia, V. (1991), 'Aboriginal Perestroika', *Arctic Circle*, vol.2, 3, pp.23-29.
Alkuperäiskansat mukana arktisten alueiden ympäristökonferenssissa (1991), *Pohjolan Sanomat*, 12.6.1991.
Alkuperäiskansojen yhteistyö tiivistynyt arktisessa ympäristönsuojelussa (1991), *Uusi Suomi*, 12.6.1991.
AMAP (1993a), *Report to Ministers on Issues of Concern to the Arctic Environment, Including Recommendations for Action. A Report from the Arctic Monitoring and Assessment Task Force (AMATF)*, AMAP Report 93:4, AMAP, Oslo.
AMAP (1993b), *Audit report: Arctic Monitoring and Assessment Programme*, AMAP Report 93:5, AMAP, Oslo.
AMAP (1995), *Guidelines for the AMAP Assessment*, AMAP Report 95:1, AMAP, Oslo.
AMAP (1996), *Minutes from the Fourth Meeting of Assessment Steering Group (ASG) of the Arctic Monitoring and Assessment Programme*, Winnipeg,

Ontario, Canada, April 14-18, 1996, AMAP Report 1996:1, AMAP, Oslo.

AMAP (1997), *Arctic Pollution Issues: A State of the Arctic Environment Report*, AMAP, Oslo.

AMEC (1997), *Information Kit.*

Andresen S., Skodvin T., Underdal A. and Wettestad J. (1994), *'Scientific' Management of the Environment? Science, Politics and Institutional Design*, CICE-RO Working Paper 94, University of Oslo, Oslo.

Andresen, S. and Wettestad, J. (1995), 'International Problem-Solving Effectiveness. The Oslo Project So Far', *International Environmental Affairs*, vol.7, 2, pp.127-149.

Archer, C. (1988), *The Soviet Union and Northern Waters*, Routledge, London and New York.

Archer, C. and Scrivener, D. (eds) (1989), *Northern Waters. Security and Resource Issues*, Croom Helm, London and Sydney.

Arctic Council Panel (1991), *To Establish Arctic Council*, CARC, Ottawa.

Arnold, R.D. (1976), *Alaska Native Land Claims*, The Alaska Native Foundation, Anchorage.

Ashley, R.K. (1986), 'The Poverty of Neoralism', in R.O. Keohane (ed.), *Neorealism and Its Critics*, Columbia University Press, New York, pp.255-300.

Assies, W.J. (1994), 'Self-Determination and the 'New Partnership'', in W.J. Assies and A.J. Hoekama (eds), *Indigenous Peoples Experiences with Self-Government. Proceedings of the Seminar on Arrangements for Self-Determination by Indigenous Peoples within National States*, 10 and 11 February 1994, Faculty of Law, University of Amsterdam, IWGIA Document 76, IWGIA, Copenhagen, pp.31-72.

Austin, J.L. (1962), *How to Do Things with Words*, Harvard University Press, Cambridge.

Barkin, S.J. and Cronin, B. (1994), 'The State and the Nation: Changing Norms and the Rules of Sovereignty in International Relations', *International Organization*, vol.48, 1, pp.107-130.

Bartelson, J. (1995), *A Genealogy of Sovereignty*, Cambridge University Press, Cambridge.

Barth, F. (1969), 'Introduction', in F. Barth (ed.), *Ethnic Groups and Boundaries: The Social Organization of Cultural Difference*, Universitetsforlaget, Oslo, pp. 9-38.

Bauman, Z. (1993), *Postmodern Ethics*, Blackwell, Oxford.

Beach, H. (1994), 'The Saami of Lapland', in *Polar Peoples: Self-Determination and Development*, Minority Rights Group, London, pp.147-205.

Behnke, A. (1993), *Structuration, Institutions and Regimes. The Case of the CSBM*, Stockholm International Studies 93:1, Stockholms Universitet, Stockholm.

Behnke, A. (1995), 'Ten Years After: The State of the Art of Regime Theory', *Cooperation and Conflict*, vol.30, 2, pp.179-197.

Bell, R.K. (1994), 'Indigenous Knowledge, Sustainable Development and Cooperative Management (The FJMC Experience)', in B.V. Hansen (ed.), *Arctic Environment. Report on the Seminar on Integration of Indigenous Peoples Knowledge*. Reykjavik, Iceland, September 1994, AEPS, Copenhagen, pp.186-201.

Berger, P.L. and Luckmann, T. (1966), *The Social Construction of Reality. A Treatise in the Sociology of Knowledge*, Doubleday & Co, New York.

Bergesen, H.O., Moe, A. and Ostreng, W. (1987), *Soviet Oil and Security Interests in the Barents Sea*, Pinter, London.

Bernauer, T. (1995), 'The Effect of International Environmental Institutions: How We Might Learn More', *International Organization*, vol.49, 2, pp.351-377.

Bernes, C. (1996), *Arktisk miljö i Norden - orörd, exploaterad, förorenad?* NORD 1996:21, Nordiska ministerrådet och Naturvårdsverket, Växjö.

Björklund, C. (1995), *A Comparison of the Legal Environmental Regimes in the Arctic and the Antarctica*, in V.G. Martin and N. Tyler (eds), Arctic Wilderness. The 5th World Wilderness Congress, Tromsø, 27th September-1st October, 1993, North American Press, Golden, pp.139-147.

Bloom, W. (1990), *Personal Identity, National Identity and International Relations*, Cambridge University Press, Cambridge.

Bonham, G.M., Jönsson, C., Persson, S. and S. Shapiro M.J. (1987), 'Cognition and International Negotiation: The Historical Discovery of Discursive Space', *Cooperation and Conflict*, vol. XXII, 1, pp.1-19.

Borgos J. (1993), 'Tradisjonell Samisk Kunnskap og Forskning', *Diedut*, 5, pp.7-21.

Boyle, A. (1996), 'The Role of International Human Rights Law in the Protection of the Environment', in A.E. Boyle and M.R. Anderson (eds), *Human Rights Approches to Environmental Protection*, Clarendon Press, Oxford, pp.43-69.

Brantenberg, O.T. (1991), 'Norway: Constructing Indigenous Self-Government in a Nation-State', in P. Jull and S. Roberts (eds), *The Challenge of Northern Regions, Northern Territory*, Australian National University, Darwin, pp.66-128.

Breyman, S. (1993), 'Knowledge as Power: Ecology Movements and Global Environmental Problems', in R.D. Lipshutz and K. Conca (eds), *The State and Social Power in Global Environmental Politics*, Columbia University Press, New York, pp.124-157.

Brock, L. (1991), 'Peace Through Parks: The Environment of the Peace Research Agenda', *Journal of Peace Research*, vol.28, 4, pp.407-423.

Bromley, D.W. (1991), *Environment and Economy. Property Rights and Public Policy*, Blackwell, Oxford.

Bröms, P., Eriksson J. and Svensson, B. (1994), *Reconstructing Survival. Evolving Perspectives on Euro-Arctic Politics*, Fritzen, Stockholm.

Bröms, P. (1995), *Environmental Security Regimes: A Critical Approach*, Research Report 23, The Swedish Institute of International Affairs, Stockholm.

Bröms, P. (1997), *Crossing the Threshols: The Forming of an Environmental Security Regime in the Arctic North*, Research Report 28, The Swedish Institute of International Affairs, Stockholm.

Brooke, L.F. (1993), *The Participation of Indigenous Peoples and the Application of their Environmental and Ecological Knowledge in the Arctic Environmental Protection Strategy*, vol. 1, ICC, Ottawa.

Brown, C. (1995), 'International Theory and International Society. The Viability of the Middle Way?', *Review of International Studies*, vol.21, 2, pp.183-196.

Brown, S., Cornell N.W., Fabian L.L., and Weiss, E.B. (1977), *Regimes for the Ocean, Outer Space, and Weather*, The Brookings Institution, Washington D.C.

Bull, H. (1977), *The Anarchical Society: A Study of Order in World Politics*, Columbia University Press, New York.

Caldwell, L.K. (1990a), *International Environmental Policy. Emergence and Dimensions*, Duke University Press, Durham and London.

Caldwell, L.K. (1990b), *Between Two Worlds. Science, the Environmental Movement, and Policy Choice*, Cambridge University Press, Cambridge.

Camilleri, J.A. and Falk, J. (1992), *The End of Sovereignty? The Politics of a Shrinking and Fragmenting World*, Edward Elgar, Aldershot.

Caporaso, J.A. (1993), 'Toward a Sociology of International Institutions: Comments on the Articles by Smouts, de Senarclens and Jönsson', *International Social Science Journal*, vol.138, pp.479-489.

Caputo, J. and Yount, M. (1993), '*Institutions, Normalization and Power'*, in J. Caputo and M. Yount (eds) Foucault and the Critique of Institutions, Pennsylvania State University, University Park, pp.3-23.

Carr, E.H. (1949), *The Moral Foundations for World Order*, The University of Denver Press.

Carr, E.H. (1951), *Twenty Years' Crisis 1919-1939. An Introduction to the Study of International Relations*, MacMillan & Co, London.

Castberg, R., Stokke, O.S., and Ostreng, W. (1994), 'The Dynamics of the Barents Region', in O.S. Stokke and O. Tunander (eds), *The Barents Region. Cooperation in the Arctic*, Sage, London, pp.71-83.

Choucri, N. (1993), 'Introduction: Theoretical, Empirical, and Policy Perspective', in N. Choucri (ed.), *Global Accord. Environmental Challenges and International Responses*, MIT Press, Cambridge, pp.1-40.

Clark, M.I. and Dryzek, J. (1987), 'The Inuit Circumpolar Conference as an International Nongovernmental Actor', in M. Stenbaek (ed.), *Arctic Policy*, Papers

Presented at the Arctic Policy Conference, September 19-21, 1985, Centre for Northern Studies and Research in Conjunction with the Inuit Circumpolar Conference and the Eben Hobson Chair, Centre for Northern Studies and Research, McGill University, Montréal, pp.215-230.

Clegg, S.R. (1989), *Frameworks of Power*, Sage, London.

Colby, M. (1990), *Environmental Management in Development. The Evolution of Paradigms*, World Bank Discussion Paper 80, World Bank, Washington D.C.

Conca, K. (1994), 'In the Name of Sustainability: Peace Studies and Environmental Discourse', in. J. Käkönen (ed.), *Green Security or Militarized Environment*, Dartmouth, Aldershot, pp.7-24.

Cortell, A.P. and Davis, J.W. Jr. (1996), 'How Do International Institutions Matter? The Domestic Impact of International Rules and Norms', *International Studies Quarterly*, vol.40,4, pp.451-478.

Creery, I. (1994), 'The Inuit (Eskimo) of Canada', in *Polar Peoples: Self-Determination and Development*, Minority Rights Group, London, pp.105-146.

Czempiel, E-O. and Rosenau, J.N. (1989), *Global Changes and Theoretical Challenges: Approaches to World Politics for the 1990s*, Lexington Books, Lexington.

Dahl, J. (1993), 'Indigenous Peoples of the Arctic', in *Arctic Challenges, in Arctic Challenges*, Report from the Nordic Council's Parliamentary Conference in Reykjavik, 16-17 August, 1993, NORD 1993:31, Nordic Council, Stockholm, pp. 103-127.

Dahl, J. (1996), 'Arctic Peoples, Their Lands and Territories', in I. Seurujärvi-Kari and U-M. Kulonen (ed.), *Essays on Indigenous Identity and Rights*, Helsinki University Press, Helsinki, pp.15-31.

Dalby, S. (1992), 'Security, Modernity, Ecology: The Dilemmas of Post-Cold War Security Discourse', *Alternatives*, vol.17, 1, pp.95-134.

Dallman, W. (1994), 'Kulturer ved Kanten av Stupet. Urbefolkninger i de russiske og sibirske norområdene', in *Polarboken 1993-94*, Norsk Polarklubb, Oslo, pp.44-62.

Davis, R.A., Richardson W.J., Thiele, L., Dietz, R. and Johansen, P. (1991), 'Report on Underwater Noise', in *The State of the Arctic Environment Report*, Arctic Centre Publications 2, Arctic Centre, Rovaniemi, pp.154-269.

Dellenbrant, J-Å. and Olsson, M-O. (1994), *The Barents Region. Security and Economic Development in the European North*, CERUM, Umeå.

Deudney, D. (1990), 'The Case Against Linking Environmental Degradation and National Security', *Millennium*, vol.19, 3, pp.461-476.

Dillon, M. (1995), 'Sovereignty and Governmentality: From the Problematics of the 'New World Order' to the Ethical Problems of the World Order', *Alternatives*, vol. 20, 3, pp.323-368.

Dosman, E. (1989), *Sovereignty and Security in the Arctic*, Routledge, London and New York.

Douglas, M. and Wildavsky A. (1983), *Risk and Culture. An Essay on the Selection of Technological and Environmental Dangers*, University of California Press, Berkeley, LA and London.

Dreyfus, H.L. and Rabinow, P. (1982), *Michel Foucault. Beyond Structuralism and Hermeneutics*, The University of Chicago Press, Chicago.

Dryzek, J. S. (1987), *Rational Ecology. Environment and Political Economy*, Basil Blackwell, Oxford.

Dryzek, J.S. (1990), *Discursive Democracy. Politics, Policy, and Political Science*, Cambridge University Press, Cambridge.

Dryzek, J.S. (1997), *The Politics of the Earth. Environmental Discourses*, Oxford University Press, Oxford.

Dunne, T. (1995a), 'The Social Construction of International Society', *European Journal of International Relations*, vol.1, 3, pp.367-389.

Dunne, T. (1995b), 'International Society. Theoretical Promises Fulfilled?', *Co-operation and Conflict*, vol.30, 2, pp.125-154.

The Earth Summit (1993), The United Nations Conference on Environment and Development (UNCED), Introduction and Commentary by Stanley P. Johnson Graham&Trotman/Martinus Nijhoff, Boston and Dordrecht.

Eckersley, R. (1992), *Environmentalism and Political Theory. Towards an Ecocentric Approach*, UCL Press, London.

Eidheim, H. (1995), 'On the Organisation of Knowledge in Sami Ethno-Politics', in T. Brantenberg, J. Hansen and H. Minde (eds), *Becoming Visible. Indigenous Politics and Self-Government*, Proceedings of the Conference on Indigenous Politics and Self-Government in Tromsø, 8-10 November 1993, The University of Tromsø, Centre for Sami Studies, Tromsø, pp.73-78.

Elliott, L.M. (1994), *International Environmental Politics. Protecting the Antarctic*. St. Martin's Press, Houndmills.

Faegteborg, M. (1993), *Towards an International Indigenous Arctic Policy (Arctic Leaders Summit)*, Arctic Information Forlag, Copenhagen.

Fairclough, N. (1992), *Discourse and Social Change*, Polity Press, Cambridge.

Fay, B. (1987), 'An Alternative View: Interpretive Social Science', in M.T. Gibbons (ed.), *Interpreting Politics*, Blackwell, Oxford, pp.82-100.

Fienup-Riordan, A. (1992), 'One Mind, Many Paths: Yup'ik Eskimo Efforts to Control Their Future', *Etudes/Inuit/Studies*, vol.16, 1-2, pp.75-83.

Finkelstein, L.S. (1995), 'What is Global Governance', *Global Governance*, vol.1, 1, pp.367-372.

Finnemore, M. (1996), 'Norms, Culture, and World Politics: Insights from Sociology's Institutionalism', *International Organization*, vol.50, 2, pp.325-347.

Flanders, N.E. (1989), 'The ANCSA Amendments of 1987 and Land Management in Alaska', *Polar Record*, vol.25, 155, pp.315-322.

Fleras, A. and Elliott, J.L. (1992), *The 'Nations Within'. Aboriginal-State Relationship in Canada, the United States and New Zealand*, Oxford University Press, Ottawa.

Florini, A. (1996), 'The Evolution of International Norms', *International Studies Quarterly*, vol.40, 3, pp.363-389.

Fondahl, G. (1995), 'The Status of Indigenous Peoples in the Russian North', *Post-Soviet Geography*, vol.36, 4, pp.215-224.

Foucault, M. (1970), *The Order of Things: An Archaeology of the Human Sciences*.Tavistock Publications, London.

Foucault, M. (1972), *Archaeology of Knowledge*, Tavistock Publications, London.

Foucault, M. (1977), 'Nietzche, Genealogy, History', in D.F. Bouchard (ed.), *Language, Counter-Memory, Practice. Selected Essays and Interviews*, Blackwell, London, pp.139-164.

Foucault, M. (1980a), 'Two Lectures', in *Power/Knowledge. Selected Interviews and Other Writings 1972-1977*, Harvester Press, Brighton, pp.78-108.

Foucault, M. (1980b), 'Afterword', in *Power/Knowledge. Selected Interviews and Other Writings 1972-1977*, Harvester Press, Brighton, pp.229-259.

Foucault, M. (1980c), 'Questions on Geography', in *Power/Knowledge. Selected Interviews and Other Writings 1972-1977*, Harvester Press, Brighton, pp. 63-77.

Foucault, M. (1982a), 'Afterword. Subject and Power', in H.L. Dreyfus and P. Rabinow (eds), *Michel Foucault. Beyond Structuralism and Hermeneutics*, The University of Chicago Press, Chicago, pp.208-226.

Foucault, M. (1982b), 'On the Genealogy of Ethics: An Overview of Work in Progress', in H.L. Dreyfus and P. Rabinow (eds), *Michel Foucault. Beyond Structuralism and Hermeneutics*, The University of Chicago Press, Chicago, pp.229-252.

Foucault, M. (1984a), 'Introduction', in M. Foucault, *The Foucault Reader* (ed. by Paul Rabinow), Pantheon Books, New York, pp.3-29.

Foucault, M. (1984b), 'Preface to the History of Sexuality, Volume II', in M. Foucault, *The Foucault Reader* (ed. by Paul Rabinow), Pantheon Books, New York, pp 333-339.

Foucault, M. (1984c), 'What is Enlightenment', in M. Foucault, *The Foucault Reader* (ed. by Paul Rabinow), Pantheon Books, New York, pp.32-50.

Foucault, M. (1988a), 'Practicing Criticism', in L.D. Kritzman (ed.), *Politics, Philosophy, Culture. Interviews and Other Writings 1977-1984*), Routledge, London and New York, pp.152-156.

Foucault, M. (1988b), 'Introduction' in L.D. Kritzman (ed.), *Politics, Philosophy, Culture. Interviews and Other Writings 1977-1984*, Routledge, London and New York, pp.ix-xxv.

Foucault, M. (1991a), 'Questions of Method', in G. Burchell, C. Gordon and P. Miller (eds), *The Foucault Effect. Studies in Governmentality. With two*

Lectures by and an interview with Michel Foucault, The University of Chicago Press, Chicago, pp.73-86.

Foucault, M. (1991b), 'Governmentality', in G. Burchell, C. Gordon and P. Miller (eds), *The Foucault Effect. Studies in Governmentality, With two Lectures by and an interview with Michel Foucault,* The University of Chicago Press, Chicago, pp.87-104.

Foucault, M. (1991c), 'Politics and the Study of Discourse', in G. Burchell, C. Gordon and P. Miller (eds), *The Foucault Effect. Studies in Governmentality. With two Lectures by and an interview with Michel Foucault,* The University of Chicago Press, Chicago, pp.53-72.

Freeman, M.M.R. and Carbyn L.N. (1988), *Traditional Knowledge and Renewable Resource Management in Northern Regions,* A Joint Publication of the IUCN Commission on Ecology and the Boreal Institute for Northern Studies, Occasional Publication 23, Boreal Institute for Northern Studies.

Freeman, M.M.R. (1992), 'Ethnoscience and Arctic Co-operation', in F. Griffiths (ed.), *Arctic Alternatives: Civility or Militarism in the Circumpolar North,* Canadian Papers in Peace Studies no. 3, Science for Peace and Samuel Stevens, Toronto, pp.79-93.

Futsaeter, G., Eidnes, G., Halmo, G., Johansen, S., Mannvik, H.P., Sydnes L.K., and Witte, U. (1991), 'Report on Oil Pollution', in *The State of the Arctic Environment Report,* Arctic Centre Publications 2, Arctic Centre, Rovaniemi, pp.270-334.

Gehring, T. (1994), *Dynamic International Regimes. Institutions for International Environmental Governance,* Peter Land, Frankfurt am Main.

Gorbachev [Gorbatsov], M. (1987), 'Mihail Gorbatsovin puhe Murmanskissa', *Sosialismin teoria ja käytäntö,* 994, 15.10.1987, pp.1-24.

Gordon, C. (1991), 'Governmental Rationality: An Introduction', in G. Burchell, C. Gordon and P. Miller (eds), *The Foucault Effect. Studies in Governmentality. With two Lectures by and an interview with Michel Foucault,* The University of Chicago Press, Chicago, pp.1-52.

Griffiths, F. (1988), 'Introduction: The Arctic as an International Political Region', in K. Möttölä (ed.), *The Arctic Challenge. Nordic and Canadian Approaches to Security and Cooperation in an Emerging International Region,* Westview Press, Boulder and London, pp.1-14.

Griffiths, F. and Young, O.R. (1989), *Sustainable Development and the Arctic. Impressions of the Co-Chairs,* Working Group on Arctic International Relations, Second Session, Ilulissat and Nuuk, Greenland 20-24 April, 1989, Report and Papers 1989-1.

Griffiths, F. (1992), 'Epilogue: Civility in the Arctic', in F. Griffiths (ed.), *Arctic Alternatives: Civility or Militarism in the Circumpolar North,* Canadian Papers in Peace Studies no. 3, Science for Peace and Samuel Stevens, Toronto, pp.279-309.

Grubb, M. et al. (1993), The 'Earth Summit' Agreements: A Guide and Assessment. An Analysis of the Rio '92 UN Conference on Environment and Development, Earthscan, London.

Haas, E.B. (1975), 'Is There a Hole in the Whole? Knowledge, Technology, Interdependence, and the Construction of International Regimes', International Organization, vol.29, 3, pp.827-876.

Haas E.B., Williams M.P. and Babai D. (1977), Scientists and World Order. The Uses of Technical Knowledge in International Relations, University of California Press, Berkeley.

Haas, E.B. (1980), 'Why Collaborate? Issue-Linkage and International Regimes', World Politics, vol. 32, 3, pp.357-405.

Haas, E.B. (1990), When Knowledge Is Power. Three Models of Change in International Organizations, University of California Press, Berkeley.

Haas, P.M. (1992a), 'Introduction: Epistemic Communities and International Policy Coordination', International Organization, vol.46, 1, pp.1-35.

Haas, P.M. (1992b), 'Banning Chlorofluorocarbons: Epistemic Efforts to Protect Stratospheric Ozone', International Organization, vol.46, 1, pp.187-224.

Haggard, S. and Simmons, B. (1987), 'Theories of International Regimes', International Organization, vol. 41, 3, pp. 491-517.

Hansen, B.V. (ed.) (1994), AEPS and Indigenous Peoples Knowledge. Report on Seminar on Integration of Indigenous Peoples Knowledge, Arctic Environmental Protection Strategy, Reykjavik, September 20-23, 1994.

Hansen, J.R., Hansson, R. and Norris, S. (eds) (1996), The State of the European Arctic Environment, EEA Environmental Monograph No. 3, European Environ ment Agency and Norwegian Polar Institute, Copenhagen.

Harris, R. (1988), Language, Saussure and Wittgenstein. How to Play Games with Words, Routledge, London and New York.

Hasenclever, A., Mayer, P. and Rittberger, V. (1996), 'Interests, Power, Knowledge: The Study of International Regimes', Mershon International Studies Review, vol. 40, suppplement 2, pp.177-228.

Heath, M. (1994), 'Birdlife International: Observer Presentation', in CAFF International Working Group Proceedings, Third Annual Meeting, Reykjavik, Iceland, September 26-28, 1994, CAFF, pp.95-101.

Heininen, L. (1991), Sotilaallisen läsnäolon riskit Arktiksessa: kohti Arktiksen säätelyjärjestelmää, Tutkimuksia 43, Tampere Peace Research Institute, Tampere.

Hekman, S.J. (1995), Moral Voices, Morals Selves. Carol Gilligan and Feminist Moral Theory, Polity Press, Cambridge.

Helander, E. (1993), 'The Role of Sami Tradition in Sustainable Development', in J. Käkönen (ed.), Politics and Sustainable Growth in the Arctic, Dartmouth, Aldershot, pp.67-80.

Helander, E. (ed) (1996), Awakened Voice. The Return of Sami Knowledge,

Diedut 4.

Hettne, B. (1994), *The Globalization of Development Theory and the Future of Development Strategies*, Padrigu Papers, Göteborg.

Hinsley, F.H. (1986), *Sovereignty*, Cambridge University Press, Cambridge.

Hjelmar, U. (1996), 'Constructivist Analysis and Movement Organizations: Conceptual Clarifications', *Acta Sociologica*, vol.39, 2, pp.169-186.

Hjorth, R. (1992), *Building International Institutions for Environmental Protection. The Case of Baltic Sea Environmental Cooperation*, Linköping University, Motala.

Hoel, A.H., Karlsen, G.R. and Breivik, A. (1993), 'Resources, Development and Environment in the Arctic', in *Arctic Challenges*, Report from the Nordic Council's Parliamentary Conference in Reykjavik, 16-17 August 1993, NORD 1993:31, Nordic Council, Stockholm, pp.65-100.

Hollis, M. and Smith, S. (1991), *Explaining and Understanding International Relations*, Clarendon Press, Oxford.

Huitfeldt T., Ries T., and Øyna, G. (1992), *Strategic Interests in the Arctic*, Forvarsstudies 4, Institutt for Forvarsstudies, Oslo.

Hunt A. and Wickham, G. (1994), *Foucault and Law: Towards a Sociology of Law as Governance*, Pluto Press, London.

Huntington, H.P. (1994), 'Traditional Ecological Knowledge of Beluga Whales: A Pilot Project in the Chuckchi and Northern Bering Seas', in B.V. Hansen (ed.), *Arctic Environment. Report on the Seminar on Integration of Indigenous Peoples Knowledge*, Reykjavik, Iceland, September 1994, AEPS, Copenhagen, pp.86-106.

Hurrell, A. (1993), 'International Society and the the Study of Regimes. A Reflective Approach', in V. Rittberger with assistance of Peter Mayer (ed.), *Regime Theory and International Theory*, Clarendon Press, Oxford, pp.49-72.

Hurrell, A. (1995), 'Explaining the Resurgence of Regionalism in World Politics', *Review of International Studies*, vol. 21, 4, pp.331-358.

Hurst, R. (1994), 'Task Force on Sustainable Development and Utilisation', in B.V. Hansen (ed.), *Arctic Environment. Report on the Seminar on Integration of Indigenous Peoples Knowledge*, Reykjavik, Iceland, September 1994, AEPS, Copenhagen, pp.121-129.

Hurwich, E. (1994), 'Circumpolar Conservation Union: Observer Presentation', in *CAFF International Working Group Proceedings*, Third Annual Meeting, Reykjavik, Iceland, September 26-28, 1994, CAFF, pp.103-104.

IASC Founding Articles (1990), IASC, Oslo.

IASC Council Meeting (1991), Report, IASC, Oslo.

IASC Council Meeting (1992), Report, IASC, Oslo.

IASC Council Meeting (1993), Report, IASC, Oslo.

IASC Council Meeting (1994), Report, IASC, Oslo.

IASC Council Meeting (1995), Report, IASC, Oslo.

IASC Meeting (1996), Report, IASC, Oslo.

ILO Convention 169 Concerning Indigenous and Tribal Peoples in Independent Countries (1989), in E. Gayuim, *The UN Draft Declaration on Indigenous Peoples. Assessment of the Draft Prepared by the Working Group on Indigenous Populations*, Juridica Lapponica 13, Institute of Northern Environmental and Minority Law, University of Lapland, Rovaniemi, pp.98-114.

Indigenous Peoples of the Soviet North (1990), IWGIA Document 67, IWGIA, Copenhagen.

Inglis, J.T. (1993), *Traditional Ecological Knowledge. Concepts and Cases*, International Program on Traditional Ecological Knowledge and International Development Research Centre, Ottawa.

IPS Information (1996), Indigenous Peoples Secretariat, Copenhagen.

James, A. (1986), *The Sovereign Statehood. The Basis of International Society*, Allen & Unwin, London.

Järvikoski, T. (1996), 'The Relation of Nature and Society in Marx and Durkheim', *Acta Sociologica*, vol.39, 1, pp.73-86.

Jensen, J. (1991), 'Report on Organochlorines', in *The State of the Arctic Environment Report*, Arctic Centre Publications 2,Arctic Centre, Rovaniemi, pp.335-384.

Jönsson, C. (ed.) (1982), *Cognitive Dynamics and International Politics*, Pinter, London.

Jönsson, C. (1993), 'Cognitive Factors in Regime Dynamics', in V. Rittberger with assistance of Peter Mayer (ed.), *Regime Theory and International Theory*. Clarendon Press, Oxford, pp.202-222.

Kaitala, S. (1994), 'World Conservation Monitoring Centre: Observer Presentation', in *CAFF International Working Group Proceedings*, Third Annual Meeting, Reykjavik, Iceland, September 26-28, 1994, CAFF, pp.91-92.

Käkönen, J. (1996), 'North Calotte as a Political Actor', in J. Käkönen (ed.), *Dreaming of Barents Region. Interpreting Cooperation in the Euro-Arctic Rim*, Research Report 73, Tampere Peace Research Institute, Tampere, pp.55-88.

Kalland, A. (1994), 'Indigenous Knowledge - Local Knowledge. Prospects and Limitations', in B.V. Hansen (ed.), *Arctic Environment. Report on the Seminar on Integration of Indigenous Peoples Knowledge*, Reykjavik, Iceland, September 1994, AEPS, Copenhagen, pp.150-167.

Kasayulie, W. (1992), 'The Self-Determination Movement of the Yupiit in Southern Alaska', *Etudes/Inuit/Studies*, vol.16, 1-2, pp.43-45.

Keeley, J.F. (1990), 'Toward a Foucauldian Analysis of International Regimes', *International Organization*, vol.44, 1, pp.83-105.

Keohane, R.O. and Nye, J.S. Jr. (1977), *Power and Interdependence. World Politics in Transition*, Little, Brown & Co., Boston and Toronto.

Keohane, R.O. (1982), 'The Demand for International Regimes', *International Organization*, vol.36, 2, pp.325-355.

Keohane, R.O. (1984), *After Hegemony: Cooperation and Discord in the World Political Economy*, Princeton University Press, Princeton NJ.

Keohane, R.O. (1988), 'International Institutions: Two Approaches', *International Studies Quaterly*, vol.32, 4, pp.379-396.

Keohane, R. O. (1989), *International Institutions and State Power. Essays in International Relations Theory*, Westview Press, Boulder.

Keohane, R.O., Haas, P.M. and Levy, M.A. (1993), 'The Effectiveness of International Environmental Institutions' in P.M. Haas, R.O.Keohane and M.A. Levy (eds), *Institutions for the Earth. Sources of Effective International Environmental Protection*, MIT, Cambridge, pp. 3-24.

Kleivan, I. (1992), 'The Arctic Peoples' Conference in Copenhagen, November 22-23, 1973', *Etudes/Inuit/Studies*, vol.16, 1-2, pp.227-236.

Klotz, A. (1995), 'Norms Reconstituting Interests: Global Racial Equality and U.S. Sanctions Against South Africa', *International Organization*, vol.49, 3, pp.451-478.

Korpijaakko-Labba, K. (1989), *Saamelaisten oikeusasemasta Ruotsi-Suomessa*, Lakimiesliiton Kustannus, Helsinki.

Korsmo, F. (1993), 'Swedish Policy and Saami Rights', *The Northern Review*, vol. 11, pp.32-55.

Koskenniemi, M. (1994), 'Mitä kansainvälinen ympäristöoikeus on? Ajatuksia ympäristönsuojelun kansainvälisistä strategioista', in. E.J. Hollo and J.K. Parkkari (eds), *Kansainvälinen ympäristöoikeus*, Gummerus, Jyväskylä.

Krasner, S.D. (1982), 'Structural Causes and Regime Consequences: Regimes as Intervening Variables', *International Organization*, vol.36, 2, pp.185-205.

Krasner, S.D. (1985), *Structural Conflict. The Third World Against Global Liberalism*,University of California Press, Los Angeles and London.

Krasner, S.D. (1989), 'Sovereignty: An Institutional Perspective', in J.A. Caporaso (ed.), *The Elusive State. International and Comparative Perspectives*, Sage, Newbury Park, pp.69-96.

Kratochwil, F. (1984), 'The Force of Prescriptions', *International Organization*, vol. 38, 4, pp.685-708.

Kratochwil, F. and Ruggie, J.G. (1986), 'International Organization: A State of the Art or an Art of the State', *International Organization*, vol.40, 4, pp.753-775.

Kratochwil, F.V. (1989), *Rules, Norms and Decisions. On the Conditions of Practical and Legal Reasoning in International Affairs and Domestic Affairs*, Cambridge University Press, Cambridge.

Kullerud, L. (1994), 'Global Resources information Database-(GRID)- Arendal', in *CAFF International Working Group Proceedings*, Third Annual Meeting, Reykjavik, Iceland. September 26-28, 1994, CAFF, pp.92-93.

Langlais, R. (1995), *Reformulating Security. A Case Study from Arctic Canada*, Humanekologiska Skrifter 13, Göteborg Universitet, Göteborg.

Larsen, F.B. (1992), 'The Quiet Life of a Revolution: Greenlandic Home Rule

1979-1992', *Etudes/Inuit/ Studies*, vol.16, 1, pp.199-226.

Lauritzen, P. (1983), *Oil and Amulets. Inuit: A People at the Top of the World* (ed. R.E. Bouchler), Breakwater Books, Canada.

Lemert, C.C. and Gillan, G. (1982), *Michel Foucault. Social Theory and Transgression*, Columbia University Press, New York.

Levy, M.A. (1993), 'Political Science and the Question of Effectiveness of International Environmental Institutions', *International Challenges*, vol.13, 2, pp.17-35.

Levy, M.A., Young O.R. and Zürn, M. (1995), 'The Study of International Regimes', *European Journal of International Relations*, vol.1, 3, pp.267-330.

Litfin, K. (1994), *Ozone Discourses. Science and Politics in Global Environmental Cooperation*, Columbia University Press, New York.

Lynch, C. (1994), 'E.H. Carr, International Relations Theory, and the Societal Origins of International Legal Norms', *International Organization*, vol. 23, 3, pp.589-619.

Lynge, A. (1992), 'Declaration of the ICC - Inuit Circumpolar Conference', *IWGIA Newsletter*, 1, pp.7-9.

Lynge, F. (1992), *Arctic Wars, Animal Rights and Endangered Peoples*, University Press of New England, Hanover.

McCormick, J. (1989), *Reclaiming Paradise. The Global Environmental Movement*, Indiana University Press, Bloomington and Indianapolis.

McKinlay, R.D. and Little, R. (1986), *Global Problems and World Order*, Pinter, London.

McNay, L. (1992), *Foucault and Feminism*, Polity Press, Cambridge.

Mähönen, O. (1997b), *Background Information on Arctic Environmental Cooperation and the Environmental Policy of the European Union*, Ympäristöministeriön moniste 18, Ympäristöministeriö, Helsinki.

Mayall, J. (1990), *Nationalism and International Society*, Cambridge University Press, Cambridge.

Melnikov, S.A. (1991), 'Report on Heavy Metals', in *The State of the Arctic Environment Report*, Arctic Centre Publications 2, Arctic Centre, Rovaniemi, pp.82-153.

Merchant, C. (1980), *The Death of Nature. Women, Ecology and the Scientific Revolution*, Harper & Row, San Francisco.

Mersheimer, J.J. (1994-5), 'The False Promise of International Institutions', *International Security*, vol.19, 3, pp.5-49.

Milner, H. (1991), 'The Assumption of Anarchy in IR Theory: A Critique', *Review of International Studies*, vol.17, 1, pp.76-85.

Milner, H. (1992), 'International Theories of Cooperation Among Nations: Strengths and Weaknesses', *World Politics*, vol.44, 3, pp.466-496.

Milner, H. (1993), 'International Regimes and World Politics: Comments on the Arcticles by Smouts, de Senarclens and Jönsson', *International Social Science*

Journal, vol.138, pp.491-497.

Minde, H. (1995), 'The International Movement of Indigenous Peoples: An Historical Perspective', in T. Brantenberg, J. Hansen and H. Minde (eds), *Becoming Visible. Indigenous Politics and Self-Government*, Proceedings of the Conference on Indigenous Politics and Self-Government in Tromsø, 8-10 November 1993, University of Tromsø, Centre for Sami Studies, Tromsø, pp.9-26.

Morehouse, T. (1989), 'Sovereignty, Tribal Government, and the Alaska Native Claims Settlement Act Amendments of 1987', *Polar Record*, vol.25, 154, pp.197-206.

Möttölä, K. (ed.) (1988), *The Arctic Challenge. Nordic and Canadian Approaches to Security and Cooperation in an Emerging International Region*, Westview Press, Boulder and London.

Myerson, G. and Rydin, Y. (1996), *The Language of Environment. A New Rhetoric*, UCL Press, London.

Myntti, K. (1995), 'The Right to Political Participation of Indigenous Peoples. The Case of the Finnish Sami', in E. Gayim and K. Myntti (eds), *Indigenous and Tribal Peoples Rights 1993 and After*, Juridica Lapponica 11, Northern Institute of Environmental and Minority Law, University of Lapland, Rovaniemi, pp.130-168.

Myntti, K. (1996), 'National Minorities, Indigenous Peoples and Various Modes of Political Participation', in F. Horn (ed.), *Minorities and their Right of Political Participation*, Juridica Lapponica 16, Northern Institute of Environmental and Minority Law, University of Lapland, Rovaniemi, pp.1-26.

Nenonen, M. (1991), 'Report on Acidification in the Arctic Countries: Man-made Acidification in a World of Natural Extremes', in *The State of the Arctic Environment Report*, Arctic Centre Publications 2, Arctic Centre, Rovaniemi, pp.7-81.

Nettheim, G. (1988), 'Peoples and Populations: Indigenous Peoples and the Rights of Peoples', in J. Crawford (ed.), *The Rights of Peoples*, Clarendon Press, Oxford, pp.107-126.

Neufeld, M. (1995), *The Restructuring of International Relations Theory*, Cambridge University Press, Cambridge.

Neumann, I.B. (1994), 'A Region-Building Approach to Northern Europe', *Review of International Studies,* vol.20, 1, pp.53-74.

Nilson, H. R. (1996), 'Environmental Factors in Multilateral Cooperation - Europeanization of the Barents Region and the Organization of Environmental Cooperation', in J. Käkönen (ed.), *Dreaming of the Barents Region. Interpreting Cooperation in the Euro-Arctic Rim*, Research Report no. 73, Tampere Peace Research Institute, Tampere, pp.251-295.

Nilson, H.R. (1997), *Arctic Environmental Protection Strategy (AEPS): Process and Organization 1991-1997. An Assessment*, Rapportserie No. 103, Norwegian Polar Institute, Oslo.

Nokkala, A. (1997) (ed.), *The European North - Challenges and Opportunities*. Ministry of Trade and Industry, Helsinki.

Northern Contaminants Program (1997), *Canadian Arctic Contaminants Assessment Report*, Indian and Northern Affairs Canada, Ottawa.

Nunavut Agreement (1993), *Agreement between the Inuit of the Nunavut Settlement Area and Her Majesty the Queen in Right of Canada*, Indian and Northern Affairs, Ottawa.

Nuttall, M. (1992), *Arctic Homeland. Kinship, Community and Development in Northwest Greenland*, University of Toronto Press, Toronto.

The Nuuk Conclusions and Recommendations on Indigenous Autonomy and Self-Government (1991), *IWGIA Newsletter*, 2, pp.9-20.

O'Meara, R.l. (1984), 'Regimes and their Implications for International Theory', *Millennium*, vol.13, 3, pp.245-264.

Onuf, N.G. (1989), *World of Our Making. Rules and Rule in Social Theory and International Relations*, University of South Carolina Press, Columbia, South Carolina.

Onuf, N.G. (1995), 'Levels', *European Journal of International Relations,* vol.1, 1, pp.35-58.

Oreskov, C. and Seijersen, F. (1993), 'Arctic: The Attack on Inuit Whale Hunting by Animal Rights Groups', *IWGIA Newsletter*, 3, pp.10-16.

Osherenko, G. and Young, O.R. (1993), 'The Formation of International Regimes: Hypotheses and Cases', in Young, O.R. and Osherenko, G. (eds) (1993), *Polar Politics. Creating International Environmental Regimes,* Cornell University Press, Ithaca and London, pp.1-21.

Osherenko, G. and Young O.R. (1989), *The Age of the Arctic. Hot Conflicts and Cold Realities*, Cambridge University Press, Cambridge.

Osherenko, G. (1995), 'Property Rights and Transformation in Russia: Institutional Change in the Far North', *Europe-Asia Studies*, vol.47, 7, pp.1077-1108.

Paakkola, O. (1991), 'Report on Radioactivity in the Arctic Region' , in *The State of the Arctic Environment Report*, Arctic Centre Publications 2, Arctic Centre, Rovaniemi, pp.385-405.

PAME (1996), *Report to the Third Ministerial Conference on the Protection of the Arctic Environment, 20-21 March 1996, Inuvik Canada*, Ministry of the Environment, Oslo.

Passmore, J. (1974), *Man's Responsibility for Nature. Ecological Problems and Western Traditions*, Charles Scribner's Sons, New York.

Peltonen, M. (1992), *Matala katse. Kirjoituksia mentaliteettien historiasta*, Hanki ja jää, Tampere.

Peluso, N.L. (1993), 'Coercive Conservation: The Politics of State Resource Control', in R.D. Lipshutz and K. Conca (eds), *The State and Social Power in Global Environmental Politics*, Columbia University Press, New York, pp.46-70.

Poelzer, G. (1995), 'Devolution, Constitutional Development and the Russian North', *Post-Soviet Geography*, vol.36, 4, pp.204-214.

Porter, G. and Brown, J.W. (1991), *Global Environmental Politics*, Westview Press, Boulder.

Price, R.E. (1982), *Legal Status of the Alaska Natives. Report to the Alaska Statehood Commission*, Department of Law, State of Alaska.

Princen, T. and Finger, M. (eds) (1994), *Environmental NGOs in World Politics. Linking the Local and the Global*, Routledge, London and New York.

Princen, T. (1994), 'NGOs: Creating a Niche in Environmental Diplomacy', in T. Princen and M. Finger (eds) (1994), *Environmental NGOs in World Politics. Linking the Local and the Global*, Routledge, London and New York, pp.29-47

Principles and Elements for a Comprehensive Arctic Policy (1992), ICC.

Prokosch, P. (1992), 'Working for the Arctic Wilderness', in T. Greiffenberg (ed.), *Sustainability in the Arctic, Proceedings from Nordic Arctic Research Forum Symposium*, Aalborg University, Aalborg.

Puchala, D.J. and Hopkins, R.F. (1982), 'International Regimes: Lessons from Inductive Analysis', *International Organization*, vol.36, 2, pp.245-275.

Report of the House of Commons Standing Committee on Foreign Affairs and International Trade (1997), *Canada and the Circumpolar World: Meeting the Challenges of Cooperation into the Twenty-First Century*, Ottawa.

Ringold, P.L. (1994), 'The Arctic Monitoring and Assessment Program', *Arctic Research of the United States*, 8, pp.101-104.

Rio Declaration on Environment and Development (1992), in *The Earth Summit. The United Nations Conference on Environment and Development* (UNCED), (Introduction and Commentary by Stanley P. Johnson), Graham&Trotman/Martius Nijhoff, London, Boston and Dordrecht, pp.117-122.

Rolston, H. III (1988), *Environmental Ethics. Duties to and Values in the Natural World*, Temple University Press, Philadelphia.

Rood, J.Q.Th. (1989), 'The Functioning of Regimes in an Interdependent World' in J.N. Rosenau and H. Tromp (eds), *Interdependence and Conflict in World Politics*, Avebury, Aldershot, pp.61-82.

Roots, E.F. (1992), 'Co-operation in Arctic Science: Background and Requirements', in F. Griffiths (ed.), *Arctic Alternatives: Civility or Militarism in the Circumpolar North*, Canadian Papers in Peace Studies no. 3, Science for Peace and Samuel Stevens, Toronto, pp.136-155.

Roots, E.F. (1993), 'The Arctic Region - Challenges and Opportunities', in *Arctic Challenges*, Report from the Nordic Council's Parliamentary Conference in Reykjavik, 16-17 August 1993, NORD 1993:31, Nordic Council, Stockholm, pp.143-154.

Rosenau, J.N. (1986), 'Before Cooperation: Hegemons, Regimes and Habit-Driven Actors in World Politics', *International Organization*, vol.40, 4, pp.849-894.

Rosenau, J.N. (1990), *Turbulence in World Politics: A Theory of Change and*

Continuity, Princeton University Press, Princeton, NJ.

Rosenau, J.N. and Czempiel, E.-O. (eds) (1992), *Governance without Government: Order and Change in World Politics*, Cambridge University Press, Cambridge.

Rothwell, D.R. (1994), 'Polar Lessons for an Arctic Regime', *Cooperation and Conflict*, vol. 29, 1, pp.55-76.

Rothwell, D.R. (1995), 'International Law and the Protection of the Arctic Environment', *International and Comparative Law Quarterly*,vol. 44, part 2, pp.280-312.

Rothwell, D.R. (1997), *The Polar Regions and the Development of International Law*. Cambridge University Press, Cambridge.

Ruggie, J.G. (1982), 'International Regimes, Transactions, and Change: Embedded Liberalism in the Postwar Economic Order', *International Organization*, vol. 36, 2, pp. 379-414.

Ruggie, J.G. (1993a), 'Territoriality and Beyond: Problematizing Modernity in International Relations', *International Organization*, vol.47, 1, pp.139-174.

Ruggie, J.G. (1993b), 'Multilateralism: The Anatomy of an Institution', in J.G. Ruggie (ed.), *Multilateralism Matters. The Theory and Praxis of an Institutional Form*. Columbia University Press, New York, pp.3-47.

Sambo, D. (1992), 'The Emerging Indigenous Human Right to Development', *IWGIA Yearbook 1991*, IWGIA, Copenhagen, pp.167-189.

Sanders, D. (1993), 'Self-Determination and Indigenous Peoples', in C. Tomuschat (ed.), *Modern Law of Self-Determination*, Martinus Nijhoff, Dordrecht, pp.55-81.

Sanghi, M. (1996), 'Interview (April 1991) with the President of the Association of the Small Peoples of the Soviet North by A. Pika', in A. Pika, J. Dahl and I.Larsen (eds), *Anxious North. Indigenous Peoples in Soviet and Post Soviet Russia*, IWGIA Document 82, IWGIA, Copenhagen, pp.65-74.

Schwartz, B. (1986), *First Principles, Second Thoughts: Aboriginal Peoples, Constitutional Reform and Canadian Statecraft*, The Institute for Research on Public Policy, Montréal.

Scrivener, D. (1996), *Environmental Cooperation in the Arctic: From Strategy to Council*, Security Policy Library No. 1/1996, The Norwegian Atlantic Committee, Oslo.

Scrivener, D. (1997), 'What Lies Ahead?', *WWF Arctic Bulletin*, 1, pp.5-6.

Searle, J.R. (1969), *Speech Acts. An Essay in the Philosophy of Language*, Cambridge University Press, Cambridge.

Senseney, R.S. (1997), 'Science and Public Policy - an Arctic Perspective', A Paper Presented to the International Arctic Science Committee, May 5, 1997, St. Petersburg, Russian Federation, in *IASC Annual and Council Meetings 1997, Reports*, IASC, Oslo.

Sergejeva, J. (1995), 'The Situation of the Sami in Kola', in E. Gayim and K. Myntti (eds), *Indigenous and Tribal Peoples Rights 1993 and After*, Juridica Lapponi ca 11, Northern Institute of Environmental and Minority Law, University of Lapland, Rovaniemi, pp.176-188.

Seurujärvi-Kari, I. (1994), 'Saamelaiset alkuperäiskansojen yhteisössä', in U.M Kulonen, J. Pentikäinen ja I. Seurujärvi-Kari (ed.), *Johdatus saamentutkimukseen*, Suomen Kirjallisuuden Seura, Helsinki, pp.171-189.

Shapiro, M.J., Bonham, G.M. and Heradstveit, D. (1988), 'A Discursive Practices Approach to Collective Decision-Making', *International Studies Quarterly*, vol.32, 4, pp.397-419.

Shapiro, M.J. (1989), 'Textualising Global Politics', in J. Der Derian and M.J. Shapiro (eds), *International/Intertextual Relations. Postmodern Readings of World Politics*, Lexington Books, Lexington, MA and Toronto, pp.11-22.

Sillanpää, L. (1994), *Political and Administrative Response to Sami Self-Determi nation. A Comparative Study of Public Administration in Fennoscandia on the Issue of Sami Land Title as an Aboriginal Right*, Comment. Scientarium Socialium 48, Societas Scientarum Fennica, Helsinki.

Simon, M. (1987), 'The Role of Inuit in International Affairs', in M. Stenbaek (ed.), *Arctic Policy*, Papers Presented at the Arctic Policy Conference, September 19-21, 1985, Centre for Northern Studies and Research in Conjunction with the Inuit Circumpolar Conference and the Eben Hobson Chair, Centre for Northern Studies and Research, McGill University, Montréal, pp.69-76.

Simons, J. (1995), *Foucault & the Political*, Routledge, London and New York.

Slezkine, Y. (1994), *Arctic Mirrors. Russia and the Small Peoples of the North*, Cornell University Press, Ithaca and London.

Smith, C. (1991), 'Indigenous Peoples and the State: Some Basic Legal Principles Seen from a Nordic Point of View', in P. Seyersted (ed.), *The Arctic: Canada and the Nordic Countries*, The Nordic Association for Canadian Studies, Lund, pp.123-128.

Smith, S.J. (1997), 'Time for a Change. Financing the Arctic Environmental Protection Strategy (AEPS) and its Working Groups', *WWF Bulletin*, vol. 4, 1, pp.8-9.

Sprout, H. and Sprout, M. (1971), *Toward a Politics of Planet Earth*, Van Nostrand Reinhold Co., New York.

Statement of Intent (1990), in *Cooperation in a Changing World,* Third Northern Regions Conference, Anchorage, Alaska.

Stavenhagen, R. (1994), 'Indigenous Rights: Some Conceptual Problems. Indigenous Peoples' Experiences with Self-Government', in W.J. Assies and A.J. Hoekama (eds), *Indigenous Peoples Experiences with Self-Government. Proceedings of the Seminar on Arrangements for Self-Determination by Indigenous Peoples within National States*. 10 and 11 February 1994, Faculty of Law, University of Amsterdam, IWGIA Document 76, IWGIA, Copenhagen, pp.9-29.

Stokke, O.S. (1990), 'The Northern Environment: Is Cooperation Coming', in M. O. Heisler (ed.), *'The Nordic Region: Changing Perspectives in International*

Relations', *The Annals of the American Academy of Political and Social Science*, Nov. 1990.

Stokke, O.S. and Tunander, O. (1994), *The Barents Region: Cooperation in Arctic Environment*, Sage, London.

Stoltenberg, T. (1992), 'The Barents Region: Reorganizing Northern Europe', *International Challenges*, vol.12, 4, pp.5-12.

Strange, S. (1982), 'Cave Hic Dragones!', *International Organization*, vol.36, 2, pp.479-496.

Suganami, H. (1983), 'The Structure of Institutionalism. An Anatomy of British Mainstream International Relations', *International Relations*, vol.7, 5, pp.2363-2381.

Suganami, H. (1989), *The Domestic Analogy and World Order Proposals*, Cambridge University Press, Cambridge.

Taylor, P. (1986), *Respect for Nature. A Theory of Environmental Ethics*, Princeton University Press, Princeton, NJ.

Taylor, C. (1987), 'Language and Human Nature', in M.T. Gibbons (ed.), *Interpreting Politics*, Blackwell, Oxford, pp.101-132.

Thornberry, P. (1991), *International Law and the Rights of Minorities*, Clarendon Press, Oxford.

Tikkanen, E. (ed.) (1995), *Kuolan saastepäästöt Lapin metsien rasitteena. Itä-Lapin metsävaurioprojektin loppuraportti*, Gummerus, Jyväskylä.

Tomasevski, K. (1995), 'Environmental Rights', in A. Eide and C. Krause (ed.), *Economic, Social and Cultural Rights. A Textbook*, Martinus Nijhoff, Dordrecht, pp.257-269.

Touraine, A. (1981), *The Voice and the Eye. An Analysis of Social Movements*, Cambridge University Press, Cambridge.

Turner, R.K. (1988), 'Sustainability, Resource Conservation and Pollution Control: An Overview', in K. Turner (ed.), *Sustainable Environmental Management: Principles and Practice*, Belhaven Press, London.

Underdal, A. (1992), 'The Concept of Regime 'Effectiveness'', *Cooperation and Conflict*, vol. 27, 3, pp.227-240.

Underdal, A. (1995), 'The Study of International Regimes', *Journal of Peace Research*, vol.32, 1, pp.113-119.

United States Announces New Policy for the Arctic Region (1995), *MAB NSN Newsletter*, p.17.

Vakhtin, N. (1992), *Native Peoples of the Russian Far North*, Minority Rights Group International Report 5, Free Press, Manchester.

Väliverronen, E. (1996), *Ympäristöuhkan anatomia. Tiede, mediat ja metsän sairaskertomus*, Vastapaino, Jyväskylä.

Vogler, J. (1996a), *The Global Commons. A Regime Analysis*, John Wiley, Chicester.

Vogler, J. (1996b), 'The Environment in International Relations: Legacies and

Contentions', in J. Vogler and M. Imber (eds), *The Environment and International Relations*, Routledge, London and New York, pp.1-21.

Waever, O. (1992), 'International Society - Theoretical Promises Unfulfilled?', *Cooperation and Conflict*, vol.27, 1, pp.97-128.

Walker, R.B.J. (1989), 'History and Structure in the Theory of International Relations', *Millennium*, vol.18, 2, pp.163-183.

Walker, R.B.J. and Mendlovitz, S.H. (1990), 'Interrogating State Sovereignty ', in R.B.J Walker and S.H. Mendlowitz (eds), *Contending Sovereignties. Redefining Political Community*, Lynne Rienner, Boulder, pp.1-12.

Walker, R.B.J. (1993), *Inside/Outside: International Relations as Political Theory*, Cambridge University Press, Cambridge.

Weiss, T.G. and Gordenker, L. (eds) (1996), *NGOs, The UN & Global Governance*, Lynne Rienner, Boulder and London.

Wendt, A.E. (1987), 'The Agent-Structure Problem in International Relations Theory', *International Organisation*, vol.41, 3, pp.335-370.

Wendt, A. and Duvall, R. (1989), 'Institutions and International Order', in E-O. Czempiel and J.O. Rosenau (eds), *Global Changes and Theoretical Challenges. Approaches to World Politics for the 1990s*, Lexington, Lexington and Toronto, pp.51-73.

Wendt, A. (1994), 'Collective Identity Formation and the International State', *American Political Science Review*, vol.88, 2, pp.384-396.

Wendt, A. (1996), 'Identity and Structural Change in International Politics', in Y. Lapid and F. Kratochwil (eds), *The Return of Culture and Identity in IR Theory*, Lynne Rienner, Boulder, pp.47-64.

Wenzel, G. (1991), *Animal Rights, Human Rights: Ecology, Economy and Ideology in the Canadian Arctic*, Belhaven, London.

Wettestad, J. (1994), 'LRTAP: The Jewel in the Crown for Discursive Diplomacy', in *Implementing Environmental Conventions*, Papers from the Second High-Level Nordic Policy Seminar. Copenhagen, Denmark October 27 and 28, 1994, Scandinavian Seminar College, Ringkjobing, pp.68-76.

Wight, M. (1992), *International Theory. The Three Traditions* (ed. by G. Wight and B. Porter), Holmes & Meier, New York.

Wilde, J. de (1991), *Saved from Oblivion: Interdependence Theory in the First half of the 20th Century. A Study on the Causality Between War and Complex Interdependence*, Dartmouth, Aldershot.

Willetts, P. (ed.) (1982), *Pressure Groups in the Global System. The Transnational Relations of Issue-Orientated Non-Governmental Organization*, Pinter, London.

Wilson, P. (1989), 'The English School of International Relations: A Reply to Sheila Grader', *Review of International Studies*, vol.15, 1, pp.49-58.

Wind, M. (1997), 'Nicholas G. Onuf: The Rules of Anarchy', in I.B. Neumann and

O. Waever (eds), *The Future of International Relations. Masters in the Making?* Routledge, London and New York, pp.236-268.

Woolin, S.S. (1988), 'On the Theory and Practice of Power', in J.Arac (ed.), *After Foucault. Humanistic Knowledge, Postmodern Challenges,* Rutgers University Press, New Brunswick and London.

World Commission on Environment and Development (1987), *Our Common Future.* Oxford University Press, Oxford.

World Conservation Strategy (1980), Prepared by the International Union for Conservation of Nature and Natural Resources (IUCN), IUCN, UNEP, WWF.

Young, O.R. (1980), 'International Regimes: Problems of Concept Formation', *World Politics,* vol.32, 3, pp.331-356.

Young, O.R. (1989), *International Cooperation. Building Regimes for Natural Resources and the Environment,* Cornell University Press, Ithaca and London.

Young, O.R. (1990), 'Global Environmental Change and International Governance', *Millennium,* vol.19, 3, pp.337-346.

Young, O.R. (1992), *Arctic Politics. Conflict and Cooperation in the Circumpolar North,* University Press of New England, Hanover and London.

Young, O.R. and Druckman, D. (eds) (1992), *Global Environmental Change. Understanding the Human Dimensions,* National Academy Press, Washington D.C.

Young, O.R. (1993b), *Public Policy and Natural Resources: Choosing Human/ Nature Relationships,* Occasional Paper Series 3, The Nelson A. Rockefeller Center for the Social Sciences at the Dartmouth College, Dartmouth.

Young, O.R. and Osherenko, G. (1993), *Polar Politics. Creating International Environmental Regimes,* Cornell University Press, Ithaca and London.

Young, O.R. and Moltke, K. von (1994), 'The Consequences of International Environmental Regimes. Report from the Barcelona Workshop', *International Environmental Affairs,* vol.6, 4, pp.348-370.

Young, O.R. (1995b), *International Governance. Protecting the Environment in a Stateless Society,* Cornell University Press, Ithaca and London.

Index

Printed and bound by CPI Group (UK) Ltd, Croydon, CR0 4YY

22/10/2024

01777626-0002